LUNENBURG COUNTY [VIRGINIA] ROAD ORDERS

1746-1764

Virginia Genealogical Society
Richmond, Virginia

Published With Permission from the

Virginia Transportation Research Council
(A Cooperative Organization Sponsored Jointly by the Virginia
Department of Transportation and
the University of Virginia)

HERITAGE BOOKS
2008

HERITAGE BOOKS
AN IMPRINT OF HERITAGE BOOKS, INC.

Books, CDs, and more—Worldwide

For our listing of thousands of titles see our website
at
www.HeritageBooks.com

Published 2008 by
HERITAGE BOOKS, INC.
Publishing Division
100 Railroad Avenue #104
Westminster, Maryland 21157

Copyright © 1993, 2004 Virginia Genealogical Society

All rights reserved. No part of this book may be reproduced or transmitted in any form or by any means, electronic or mechanical, including photocopying, recording or by any information storage and retrieval system without written permission from the author, except for the inclusion of brief quotations in a review.

International Standard Book Number: 978-0-7884-3669-7

LUNENBURG COUNTY ROAD ORDERS 1746-1764

by

Nathaniel Mason Pawlett
Faculty Research Historian

and

Tyler Jefferson Boyd
Research Assistant

(The opinions, findings, and conclusions expressed in this report are those of the authors and not necessarily those of the sponsoring agencies.)

Virginia Transportation Research Council
(A Cooperative Organization Sponsored Jointly by the Virginia
Department of Transportation and
the University of Virginia)

Charlottesville, Virginia

January 1993
Revised April 2004
VTRC 93-R17

Historic Roads of Virginia

Louisa County Road Orders 1742-1748, by Nathaniel Mason Pawlett. 57 pages, indexed, map.

Goochland County Road Orders 1728-1744, by Nathaniel Mason Pawlett. 120 pages, indexed, map.

Albemarle County Road Orders 1744-1748, by Nathaniel Mason Pawlett. 52 pages, indexed, map.

The Route of the Three Notch'd Road, by Nathaniel Mason Pawlett and Howard Newlon, Jr. 26 pages, illustrated, 2 maps.

An Index to Roads in the Albemarle County Surveyors Books 1744-1853, by Nathaniel Mason Pawlett. 10 pages, map.

A Brief History of the Staunton and James River Turnpike, by Douglas Young, 22 pages, illustrated, map.

Albemarle County Road Orders 1783-1816, by Nathaniel Mason Pawlett. 421 pages, indexed.

A Brief History of the Roads of Virginia 1607-1840, by Nathaniel Mason Pawlett. 41 pages, 3 maps.

A Guide to the Preparation of County Road Histories, by Nathaniel Mason Pawlett. 26 pages, 2 maps.

Early Road Location: Key to Discovering Historic Resources? By Nathaniel Mason Pawlett and K. Edward Lay. 47 pages, illustrated, 3 maps.

Albemarle County Roads 1725-1816, by Nathaniel Mason Pawlett. 98 pages, illustrated, 5 maps.

Backsights: A Bibliography, by Nathaniel Mason Pawlett. 29 pages.

Orange County Road Orders 1734-1749, by Ann Brush Miller. 323 pages, indexed, map.

Spotsylvania County Road Orders 1722-1734, by Nathaniel Mason Pawlett. 152 pages, indexed, map.

Brunswick County Road Orders 1732-1746, by Nathaniel Mason Pawlett. 81 pages, indexed, map.

Orange County Road Orders 1750-1800, by Ann Brush Miller. 394 pages, indexed, map.

Library of Congress Catalogue Card
No. 91-06777

A Note on the Methods, Editing and Dating System

The road and bridge orders contained in the order books of an early Virginia county are the primary source of information for the study of its roads. When extracted, indexed and published by the Virginia Transportation Research Council, they greatly facilitate this. All of the early county court order books are in manuscript, sometimes so damaged and faded as to be almost indecipherable. Usually rendered in the rather ornate copperplate script of the time, the phonetic spellings of this period often serve to further complicate matters for the researcher and recorder.

With these road orders available in an indexed and cross-referenced published form, it will be possible to produce chronological chains of road orders illustrating the development of many of the early roads of a vast area from the threshold of settlement up through the middle of the eighteenth century. Immediate corroboration for these chains of early road orders will usually be provided by other evidence such as deeds, plats and the Confederate Engineers maps. Often, in fact, the principal roads will be found to survive in place under their early names.

With regard to the general editorial principles of the project, it has been our perception over the years as the road orders of Louisa, Hanover, Goochland and Albemarle have been examined and recorded that road orders themselves are really a variety of "notes", often cryptic, incomplete or based on assumptions concerning the level of knowledge of the reader. As such, any further abstracting or compression of them would tend to produce "notes" taken from "notes", making them even less comprehensible. The tendency has therefore been in the direction of restraint in editing, leaving any conclusions with regard to meaning up to the individual reader or researcher using these publications. In pursuing this course, we have attempted to present the reader with a typescript text, which is as near a type facsimile of the manuscript itself as we can come.

Our objective is to produce a text that conveys as near the precise form of the original as we can, reproducing all the peculiarities of the eighteenth-century orthography. While some compromises have had to be made due to the keyboard of the modern typewriter, this was really not that difficult a task. Most of their symbols can be accommodated by modern typography, and most abbreviations are fairly clear as to meaning.

Punctuation may appear misleading at times, with unnecessary commas or commas placed where periods should be located; appropriate terminal punctuation is often missing or else takes the form of a symbol such as a long dash, etc. The original capitalisation has been retained insofar as it was possible to determine from the original manuscript whether capitals were intended. No capitals have been inserted in place of those originally omitted. The original spelling and syntax have been retained throughout, even including the obvious errors in various places, such as repetitions of words and simple clerical errors. Ampersands have been retained throughout to include such forms as "&c.^{a}" for "etc." Superscript letters have also been retained where used in y^e., y^t., s^d. The thorn symbol (y), pronounced as "th," has been retained in the aforesaid "y^e.", pronounced "the", and "y^t." (that), along with the tailed p which the limitations of the modern typewriter have forced us to render as a capital "p". This should be taken to mean either "per" (by), "pre" or "pro" (sometimes "par" as in "Pish" for parish) as the context by the order may demand. For damaged and missing portions of the manuscript we have used square brackets to denote the [missing], [torn] or [illegible] portions. Due to the large number of ancient forms of spelling, grammar and syntax, it has been deemed impracticable to insert the form [sic] after each one to indicate a literal rendering. Therefore, the reader must assume that apparent errors are merely the result of our literal transcription of the road orders, barring the introduction of typographical errors of course. If in any case this appears to present insuperable problems, resort should be made to the original records available for examination at Lunenburg Court House.

As to dating, most historians and genealogists who have worked with early Virginian records will be aware of the English dating system in use down to 1752. Although there was an eleven-day difference from our calendar in the day of the month, the principal difference lay in the fact that the beginning of the year was dated from March 25 rather than January 1, as was the case from 1752 onward to the present. Thus January, February and March (to the 25th) were the last three months in a given year and the new year came in only on March 25.

Early Virginian records usually follow this practice, though in some cases dates during these three months will be shown in the form 1732/3, showing both the English date and that in use on the Continent, where the year began January 1. For researchers using material with dates in the English style, it is important to remember that under this system (for instance) a man might die in January 1734 yet convey property or serve in public office in June 1734, since June came before January in a given year under this system.

LUNENBURG COUNTY ROAD ORDERS 1746-1764

by

Nathaniel Mason Pawlett
Faculty Research Historian

and

Tyler Jefferson Boyd
Research Assistant

INTRODUCTION

The roads are under the government of the county courts, subject to be controuled by the general court. They order new roads to be opened whenever they think them necessary. The inhabitants of the county are by them laid off into precincts, to each of which they allot a convenient portion of the public roads to be kept in repair. Such bridges as may be built without the assistance of artificers, they are to build built. If the stream be such as to require a bridge of regular workmanship, the court employs workmen to build it, at the expense of the whole county. If it be too great for the county, application is made to the general assembly, who authorize individuals to build it, and to take a fixed toll from all passengers, or give sanction to such other proposition as to them appears reasonable.
Thomas Jefferson, *Notes on the State of Virginia*, 1781.

The establishment and maintenance of public roads was one of the most important functions of the County Court during the colonial period in Virginia. Each road was opened and maintained by an Overseer of Highways appointed by the Gentlemen Justices yearly. He was usually assigned all the "Labouring Male Titheables" living on or near the road for this purpose. These individuals then furnished all their own tools, wagons, and teams and were required to labour for six days each year on the roads.

Major projects, such as bridges over rivers, demanding considerable expenditures were executed by Commissioners appointed by the Court to select the site and to contract with workmen for the construction. Where bridges connected two counties, a commission was appointed by each and they cooperated in executing the work.

At its inception, Brunswick County comprised a large part of the Piedmont frontier east of the Blue Ridge and south of the James River, the as yet unsettled lands stretching westward to the Blue Ridge. Its distant boundary was intended both to facilitate western settlement and to increase the county's effectiveness as a buffer against the French influence advancing into the center of the continent.

From 1732 to 1746, Brunswick was a giant parent county; by the end of this time, it had shrunk to very nearly its present size, with Lunenburg taking its place. The scale of this county also made administration unwieldy, and like other large frontier counties created as a response to continued westward movement, Lunenburg began to lose its territory within six years of its creation.

The road orders contained in this volume cover the period from 1746, when Lunenburg's county government first became operational, through the creation of Halifax in 1752 and Bedford in 1754, down to the creation of Charlotte and Mecklenburg from it in 1765. As such, they are the principal extant evidence concerning the early development of roads over a vast area of Southside Virginia stretching into the very shadows of the Blue Ridge to include Pittsylvania, Henry, Franklin and Patrick, as well as major portions of Appomattox, Bedford and Campbell counties.

Note: In the 2004 digitizing of this volume, a number of typographic errors were discovered in the index. These have been corrected and the index slightly reformatted from the original.

THE DEVELOPMENT OF LUNENBURG COUNTY

Note: As originally published in paper format, this volume included maps showing the evolution of the county. Maps are not included in the revised/electronic version due to legibility and file size considerations. Instead, a verbal description is provided.

As originally formed (created in 1720, and with its county government established in 1732), Brunswick County, the parent county of Lunenburg, covered a large portion of Southside Virginia. Its boundaries extended eastward as far as modern Southampton, westward to the Blue Ridge Mountains, southward to North Carolina, and northward to Botetourt County.

Lunenburg County was created from Brunswick County in 1746, and included within its boundaries most of its parent's former territory. Brunswick was reduced to a fraction of its former size (comprising only modern day Brunswick and Greensville counties). Lunenburg included the modern counties of Lunenburg, Mecklenburg, Charlotte, Halifax, Pittsylvania, Henry, Patrick and Franklin, as well as portions of Bedford, Campbell, and Appomattox counties. As settlement proceeded westward, the separation of Halifax, Mecklenburg, Bedford and Charlotte counties between 1752 and 1765 brought Lunenburg County to its present boundaries.

Lunenburg County Road Orders

5 May 1746 Old Style, Page 6
James Easter is Appointed Surveyor of the Road from the Mouth of Ash Camp Creek the most Convenient way into Colo. Randolph's Road, and it is Ordered that Thomas Jones's Male labouring Tithables, Clement Read's Male Labouring Tithables and Philip Jones's Male Labouring Tithables assist in clearing the same.

5 May 1746 Old Style, Page 6
Lewis Delony Gent. is appointed Surveyor of the River Road from the dividing Line up to Allen's Creek And it is Ordered that all Male Labouring Tithable Persons Convenient thereto assist in clearing the same.

5 May 1746 Old Style, Page 6
William Howard Gent. is appointed Surveyor of the Road from Allen's Creek to Butcher's Creek, And it is Ordered that all Male Labouring Tithable Persons Convenient thereto assist in Clearing the same.

5 May 1746 Old Style, Page 6
Abraham Cooke Gent. is appointed Surveyor of the Road from Butcher's Creek to Blew Stone And it is Ordered that all Male Labouring Tithable Persons Convenient thereto assist in Clearing the same.

5 May 1746 Old Style, Page 6
William Harris is Appointed Surveyor of the Road from Blew Stone to Cargills Ferry on Staunton River And it is Ordered that all Male Labouring Tithable Persons convenient thereto assist in Clearing the same.

5 May 1746 Old Style, Page 7
Richard Womack is appointed Surveyor of a Road to be Cleared from the middle Fork of Little Roanoake into Falling River Road, And it is Ordered that all the Male Labouring Tithable Persons belonging to Thomas Spencer Joseph Morton Richard Womack and Robert Childress assist in Clearing the same.

2 June 1746 Old Style, Page 13
John Mead is Appointed Overseer of the Road from Otter River to the North end of the long Mountains at Poplar Spring And it is Ordered that all the Male Labouring Tithable Persons Convenient thereto assist in clearing the same.

2 June 1746 Old Style, Page 13
John Beard is Appointed overseer of the same Road from the Poplar Spring to his Road at the head of Appromattox river And it is Ordered that all the Male Labouring Tithable Persons convenient thereto assist in clearing the same.

2 June 1746, Old Style 13
Robert Baber is Appointed Overseer of a Road to be Cleared from Christopher Irvins Ford on Otter River to William Calloway's Mill the most convenient way and it is Ordered that all the Male Labouring Tithable Persons convenient thereto assist in clearing the same.

2 June 1746 Old Style, Page 13
Ordered that John Nance open and clear the Road from Winninghams Ford on Maherrin River to Randolphs Road on Little Roenoke as Twittys Path now goes, And it is Ordered that the said Nance with all Male Labouring Tithable Persons convenient thereto clear the same According to Law --

2 June 1746 Old Style, Page 14
Young Stokes is Appointed Overseer of the Road from Winninghams Foard on Maherrin River to Nottoway River And it is Ordered that all the Male Labouring Tithable Persons convenient thereto assist in Clearing the same.

2 June 1746 Old Style, Page 14
Thomas Bouldin is Appointed Overseer of the Road from the Great Lick on Little Roenoke to Nances Road And it is Ordered that all the Male Labouring Tithable Persons convenient thereto Assist in clearing the same.

1 September 1746 Old Style, Page 52
On the Petition of Timothy Dalton and others it is Ordered that a Road be Cleared from Otter River at the Fish Dam Ford to Snow Creek the best and most convenient way, and the said Timothy Dalton is appointed

Surveyor thereof And it is Ordered that he with all the Male Labouring Tithables convenient to the said Road forthwith Clear and keep the same in repair from William Verdemans upwards.

1 September 1746 Old Style, Page 63
On the Motion of John Caldwell Gent. and For Reasons appearing to the Court leave is given him to lay off and Divide the Cubb Creek Road into four Equal parts And it is Ordered that he apportion and set apart the Number of Male labouring Tithables belonging to the said Road in such manner as he shall think fit.

1 September 1746 Old Style, Page 64
On the Motion of Lewis Delony Gent. it is Ordered that a Road be laid off and cleared from this Courthouse into Hogins's Road the best and most Convenient way. And John Humphries is appointed Surveyor thereof And it is Ordered that he with all the Male labouring Tithables convenient to the said Road forthwith Clear and keep the same in Repair according to Law.

1 September 1746 Old Style, Page 64
On the Motion of Leeways Delony Gent. it is Ordered that a Road be laid off and cleared from the place where the new Road whereof John Humphries is Surveyor Intersects Hogins's Road to the Church, the best and most convenient way and the said Lewis Delony is Appointed Surveyor thereof And it is Ordered that he with all the Male labouring Tithables convenient to the said Road forthwith Clear and keep the same in Repair According to Law.

1 September 1746 Old Style, Page 65
On the Motion of Abraham Martin it is Ordered that a Road be laid off and cleared the best and most Convenient way from this Courthouse to Joseph Morton's Mill on Little Ronoke and the said Abraham Martin is appointed Surveyor of that part thereof as leads into Kings Road And it is Ordered that he with all the Male labouring Tithables convenient to the said Road forthwith Clear and keep the same in Repair according to Law.

1 September 1746 Old Style, Page 65
On the Motion of Abraham Martin it is Ordered that a Road be laid off and Cleared. [Large blank in Book]

1 September 1746 O. S., Page 65
On the Motion of Cornelius Cargill Gent. it is Ordered that a Road be laid off and Cleared the best and most Convenient way from Cargills Ferry to this Courthouse And John Cargill is appointed Surveyor thereof And it is Ordered that he with all the Male Labouring Tithables convenient to the said Road forthwith Clear and keep the same in Repair according to Law.

1 September 1746 O. S., Page 66
On the Motion of Thomas Lanear Gent. It is Ordered that a Road be laid off and cleared the best and most Convenient Way from Mitchells Ferry to this Courthouse and the said Thomas Lanear is appointed Surveyor thereof And it is Ordered that the said Lanear with all the Male Labouring Tithables Convenient to the said Road forthwith Clear and keep the same in Repair according to Law.

1 September 1746 O. S., Page 67
On the Motion of Mathew Talbot Gent. it is Ordered that a Road be laid off and cleared the best and most Convenient way from the Forks of Seneca where Dunnahoe lives to the Mouth of Falling River and William Owen is appointed Surveyor thereof And it is Ordered that he with all the Male Labouring Tithable Persons from the Mouth on the upper side of Falling and on both sides of Stanton to the Mouth of Otter River lay open, Clear and keep the same in Repair according to Law.

6 October 1746 O. S., Page 78
On the Motion of James Arnold he is Appointed Surveyor of the Road leading from Mountain Creek to Buckhorn upwards and from the Mouth of the Creek to the Extent of the County downwards to John Parkers' And it is Ordered that the said Arnold with all the Male Labouring Tithable Persons convenient to the said Road forthwith Clear and keep same in Repair according to Law.

6 October 1746 O. S., Page 78
On the Motion of Richard Fox it is Ordered that a Road be laid off and Cleared from the Horse Ford on Ronoke to Kings Ford on the South side of Ronoke and the said Richard Fox is appointed Surveyor thereof And it is Ordered that the said Fox with all the Male Labouring Tithable Persons Convenient to the said Road forthwith Clear and keep the same in Repair According to Law.

6 October 1746 O. S., Page 78
On the Motion of John Gilliam it is Ordered that a Road be laid off and cleared from Mitchells Ford to Kings Ford and from thence to the Church And the said John Gilliam is appointed Surveyor thereof And it is Ordered that the said Gilliam with all the Male Labouring Tithable Persons Convenient to the said Road forthwith Clear and keep the same in Repair according to Law.

6 October 1746 O. S., Page 78
On the Motion of Lewis Delony Gent. it is Ordered that a Road be laid off and cleared from Hickmans Ferry to Miles's Creek Bridge and Joseph Hickman is Appointed Surveyor thereof And it is Ordered that the said Hickman with all the Male Labouring Tithable Persons convenient to the said Road forthwith Clear and keep the same in Repair according to Law.

6 October 1746 O. S., Page 79
On the Motion of Robert Jones Gent. it is Ordered that a Road be laid off and Cleared from this Courthouse to Youngs Mill and the Buffelo Lick and Abraham Martin is appointed Surveyor thereof And it is Ordered that the said Martin with all the Male Labouring Tithable Persons convenient to the said Road forthwith Clear and keep the same in Repair according to Law.

1 December 1746 O. S., Page 81
Ordered That for the future it be a standing Rule with this Court that no Motions or Petitions for Roads shall be heard and Recorded except only in the Months of October and March Yearly.

2 February 1746 O. S., Page 109
On the Motion of William Jones and for Reasons appearing to the Court leave is given him to turn the Road leading near his Plantation whereof Thomas Lanear Gent. is surveyor.

2 March 1746 O. S., Page 115
On the Motion of Thomas Twitty Lycense is granted him to keep Ferry over the Horse Ford on Ronoke River in this County on his giving Bond and Security according to Law, and the Court doth set the Rates of the said Ferry as follows, to wit, for a Man four pence and for a Horse the same.

3 March 1746 O. S., Page 129
Upon the Petition of John Lucas leave is given him to lay open and Clear a Road according to the prayer of his Petition and the said John Lucas is appointed Surveyor thereof And it is Ordered that he with all the Male Labouring Tithables Convenient to the said Road do forthwith Clear and keep the same in Repair According to Law.

3 March 1746 O. S., Page 129
Ordered that a Road be laid off and Cleared from Scott's fford on Maherrin River, the best and most Convenient way to this Courthouse, and John Brown is appointed Surveyor thereof And it is Ordered that the said Brown with all the Male Labouring Tithables Convenient to the said Road do forthwith Clear and keep the same in Repair according to Law.

3 March 1746 O. S., Page 130
Robert Henry Dyer, Lazarus Williams, and William Hankins are appointed Surveyors of the Road leading from the ffork of Nottoway to Maherrin River in this County and they are to divide the same in Equal Districts between themselves, And it is Ordered that they and each of them with all the Male Labouring Tithable Persons convenient to their respective Districts forthwith Clear and keep the same in Repair According to Law.

3 March 1746 O. S., Page 130
Ordered that a Road be laid off and Cleared from William Bean's on Dan River the best and most Convenient way to Banister from thence the best and most Convenient way to the North River at Cargill's Horse Ford, and from thence the best and most Convenient Way to this Courthouse And Peter Wilson is appointed Surveyor of that part of the said Road which leads from the said Beans to Sandy River and it is Ordered that the said Wilson with all the Male Labouring Tithable Persons Convenient to the said Road do forthwith Clear and keep the same in Repair according to Law.

3 March 1746 O. S., Page 130
William Hogan is Appointed Surveyor of the Road leading from Sandy River to the Double Creeks And it is Ordered that the said Hogan with all the Male Labouring Tithable Persons Convenient to the said Road forthwith Clear and keep the Same in Repair according to Law.

3 March 1746 O. S., Page 131
Jeremiah Viditto is Appointed Surveyor of the Road leading from the Double Creeks to Banister And it is Ordered that the said Viditto with all the Male Labouring Tithable Persons Convenient to the said Road forthwith Clear and keep the same in Repair according to Law.

3 March 1746 O. S., Page 131
William Wynne is Appointed Surveyor of the Road leading from Banister to the North River at Cargill's Horse fford, and it is Ordered that the said Wynne with all the Male Labouring Tithable Persons Convenient to the said Road forthwith Clear and keep the same in Repair according to Law.

3 March 1746 O. S., Page 131
John Cargill is Appointed Surveyor of the Road leading from the North River at Cargill's Horse fford to this Courthouse And it is Ordered that the said John Cargill with all the Male Labouring Tithable Persons Convenient to the said Road forthwith Clear and keep the same in Repair according to Law.

3 March 1746 O. S., Page 132
David Stokes and Lyddall Bacon Gent. are appointed to treat with and Solicit the Justices of the County Court of Amelia for their Concurrence in Building a Bridge over Nottoway River where Willingham's Road crosses the same and that they Report the Sentiments of that Court thereupon to this Court.

3 March 1746 O. S., Page 132
On the Motion of Abraham Cooke Gent. and for Reasons appearing to the Court It is Ordered that the Road whereof he the said Abraham Surveyor be Extended and Cleared to this Courthouse And it is further Ordered that the said Cooke with all the Male Labouring Tithable Persons convenient to the said Road forthwith Clear and keep the same in Repair according to Law.

3 March 1746 O. S., Page 132
Ordered that all the Male Labouring Tithable Persons on Twittys Creek, and between the Great Lick and Ash Camp be aded to the Gang belonging to the Road whereof Thomas Bouldin Gent. is Surveyor, and that they forthwith assist in Clearing and keeping the same in Repair according to Law.

3 March 1746 O. S., Page 132
Upon the Petition of Peter Hudson it is Ordered that a Road be laid off and cleared the best and most Convenient way from Graves's fford to Scotts fford on Maherrin River and the said Peter Hudson is appointed Surveyor thereof And it is Ordered that he with all the Male Labouring Tithable Persons Convenient to the said forthwith Clear and keep the same in Repair according to Law.

3 March 1746 O. S., Page 132
On the Motion of Lewis Delony Gent. ffield Jefferson is appointed Surveyor of the Road leading from Allen's Creek Bridge to Miles's Creek Bridge and it is Ordered that the said Jefferson with all the Male Labouring Tithable Persons convenient to the said Road forthwith Clear and keep the same in Repair According to Law.

3 March 1746 O. S., Page 132
On the Motion of Lewis Delony Gent. Samuel Manning is Appointed Surveyor of the Road, leading from Miles's Creek Bridge to the Extent of the County downwards and it is Ordered that the said Manning with all the Male Labouring Tithable Persons Convenient to the said road forthwith Clear and keep the same in Repair according to Law.

3 March 1746 O. S., Page 133
Mackness Goode is Appointed Surveyor of the Road leading from this Courthouse to the Woolf Pit near Kings Road and it is Ordered that the said Goode forthwith with all the Male Labouring Tithable Persons Convenient to the said Road Clear and keep the same in Repair according to Law.

3 March 1746 O. S., Page 133
Abraham Martin is Appointed Surveyor of the Road leading from the Woolf Pit near Kings Road to Randolph's Road, and it Ordered the said Martin with all the Male Labouring Tithable Persons Convenient to the said Road forthwith Clear and keep the same in Repair according to Law.

3 March 1746 O. S., Page 133
John Young is Appointed Surveyor of the Road leading from his Mill to Randolph's Road, and it is Ordered that the said Young with all the Male Labouring Tithable Persons convenient to the said Road forthwith Clear and keep the same in Repair according to Law.

3 March 1746 O. S., Page 133
On the Motion of John Twitty it is Ordered that a Road be laid off and Cleared the best and most Convenient way from this Courthouse to Lucas's Road and the said Twitty is Appointed Surveyor thereof And it is Ordered that the said Twitty forthwith with all the Male Labouring Tithable Persons Convenient to the said Road clear and keep the same in Repair According to Law.

6 April 1747 O. S., Page 158
On the Motion of Field Jefferson Gent. License is granted him to keep Ferry over Ronoke River from his Landing to Hampton's provided he give Bond and Security for the same According to Law and he is allowed Three pence three farthings for a Man and the same for a Horse.

7 September 1747 O. S., Page 286
This Court from many Months Experience taking into their Consideration the many Grievances that attend the present Situation of this Court-house, which are as follows to wit, "That the Water near and Convenient and which is now made use of is unclean, unwholesome, very bad and not fit to drink: that the place where the Courthouse is Scituate is not Centrical but Inconvenient to the Majority of the Inhabitants of this County And is nearer the Country line than the Line which divides this from Brunswick County by about Ten Miles and is so illy Scituated that it is Impracticable to have Convenient and necessary Roads to lead to it from hardly any part of the County It is therefore Ordered that John Hall, David Stokes and Clement Reed Gentlemen do Represent to the Honourable the Governor and Council of this Colony the several Grievances aforesaid in Order to have them redressed, And to obtain such an Order as will be for the Ease and Conveniency of the Inhabitants of this County.

5 October 1747 O. S., Page 291
Upon the Petition of Richard Booker on behalf of himself and the Inhabitants of this County it is Ordered that a Road be laid open and Cleared from Mays's fford on Stanton River to Turnys Creek, thence down the nearest and best way to Cunninghams fford on Cubb Creek And from thence down the Road that was formerly Ordered by Brunswick Court to Cobbs Road And it is Ordered that all the Male Labouring Tithable Persons convenient to the said Road (Except William Gill) Assist in Clearing the same.

2 November 1747 O. S., Page 295
On the Motion of Field Jefferson it is Ordered that a Road be laid off and Cleared out of Delony's Road to the said Jefferson's Ferry, and from thence towards the place where the Court for the County of Granville in the Province of North Carolina is intended to be held and the said ffield Jefferson is Appointed Surveyor thereof And it is Ordered that he with all the Male Labouring Tithable Persons Convenient to the said Road forthwith Clear and keep the same in Repair According to Law.

2 November 1747 O. S., Page 295
On the Motion of Abraham Martin it is Ordered that a Road be laid off and Cleared the best and most Convenient way from Young's Mill to this Court House and Mackerness Goode is Appointed Surveyor of that part thereof as leads from this Courthouse to Kings Road And it is Ordered that the said Goode with all the Male Labouring Tithable Persons convenient to the said Road forthwith Clear and keep the same in Repair According to Law.

2 November 1747 O. S., Page 296
Abraham Martin is Appointed Surveyor of the New Roads leading from Kings Road to Randolph's Road //-//-//-//-- And it is Ordered that he with all the Male Labouring Tithable Persons convenient to the said Road forthwith Clear and keep the same in Repair According to Law.

2 November 1747 O. S., Page 296
John Young is Appointed Surveyor of the new Road leding from his the said Young's Mill to Randolph's Road And it is Ordered that he with all the Male Labouring Tithable Persons convenient to the said Road forthwith Clear and keep the same in Repair According to Law.

2 November 1747 O. S., Page 296
On the Motion of John Mead it is Ordered that a Road be laid off and Cleared the best and most Convenient way from the ffish Dam on Otter River to Goose Creek at a fford called Shorts fford and from thence to Stanton River a little below Nicholas Hayle's and from thence to Black Water and William Morgan is Appointed Surveyor of that part thereof as leads from the ffish Dam to Stanton River aforesaid And it is Ordered that the said Morgan with all the Male Labouring Tithable Persons Convenient to the said Road forthwith Clear and keep the same in Repair According to Law.

2 November 1747 O. S., Page 296
Robert Jones is Appointed Surveyor of the New Road leading from Stanton River a little below Nicholas Hayle's to Black Water And it is Ordered that the said Jones with all the Male Labouring Tithable Persons convenient to the said Road forthwith Clear and keep the same in Repair According to Law.

2 November 1747 O. S., Page 297
On the Motion of Abraham Cooke Gent. It is Ordered that the said Cook together with William Tabor lay off and Mark the best and most Convenient way for a Road from Blew Stone to Maherrin below the ffork and make Report thereof here to the next Court.

2 November 1747 O. S., Page 297
Samuel Wilson is Appointed Surveyor of the Road leading from the North Maherrin into Abraham Cock's Road And it is Ordered that the said Wilson with all the Male Labouring Tithable Persons convenient to the said Road forthwith Clear and keep the same in Repair According to Law.

2 November 1747 O. S., Page 297
Philemon Russell is Appointed Surveyor of the Road leading from flat Road to the County line downwards And it is Ordered that the said Russell with all the Male Labouring Tithable Persons convenient to the said Road forthwith Clear and keep the same in Repair According to Law.

7 December 1747 O. S., Page 324
County Levy
To William Tabor for four Days Attendance by Order of Court in Marking out a Road
12 (shillings)

8 December 1747 O. S., Page 327
Hugh Moore is appointed Surveyor of the Road leading from the Double Creeks to the Middle fork of Miery Creek, And it is Ordered that all the Male Labouring Tithable Persons Convenient thereto, Assist in Clearing and keeping the same in Repair.

8 December 1747 O. S., Page 327
Joseph Moore is appointed Surveyor of the Road Leading from the Middle fork of Miery Creek to the Horse Ford on Banister River, And it is Ordered that all the Male Labouring Tithable Persons Convenient thereto, Assist in Clearing and Keeping the same in Repair.

1 February 1747 O. S., Page 387
On the Motion of Adlard David Lycence is granted him to keep a Ferry in his house Provided he give Bond and Security Acording to Law in the Clerks Office and he is Allowed Seven Pence half penny for a Man and Horse.

7 March 1747 O. S., Page 388
Upon the Petition of Luke Smith it is Ordered that a Road be laid off and Cleard the best and most Convenient way from Aarons Creek to Robert Mitchels Ford and the said Luke Smith is Appointed Surveyor thereof, And it is Order'd that all the Male Labouring Tithable Persons Convenient thereto Assist in Clearing the same.

7 March 1747 O. S., Page 391
Benjamin Harrison is appointed Surveyor of the Road leading from Twittys Ferry to the County line and tis Ordered that he with the Male Labouring Tithable Persons following, to wit, William Murphey, William Douglass, Henry Rottenburry, Jeremiah Ellis & Barnaby Murphey and their Male Tithables and all others Convenient to the said Road forthwith Clear and keep the same in Repair according to Law.

7 March 1747 O. S., Page 391
Richard Fox is appointed Surveyor of the Road leading from John Davis's to Twittys Ferry and so along the New Road leading to Granville Court in the Province of North Carolina, and 'tis Ordered that all the Male Labouring Tithable Persons Convenient thereto Assist in keeping the same in Repair.

7 March 1747 O. S., Page 393
Ordered that a Road be laid off and Clear'd the best and most Convenient way from David Logins to Youngs Mill Road, from thence to Charles Talbots and from thence to the Mossing ford, and David Logan is appointed Surveyor thereof, And 'tis Ordered that all the Male Labouring Tithable Persons Convenient thereto Assist in Clearing the same.

7 March 1747 O. S., Page 393
William Hawkins is appointed Surveyor of the Road leading from the North River to Reedy Creek Bridge, & 'tis Ordered that all the Male Labouring Tithable Persons Convenient thereto Assist in keeping the same in Repair.

7 March 1747 O. S., Page 393
Lazarus Williams is appointed Surveyor of the Road Leading from Reedy Creek Bridge to Diers Pine, and 'tis Ordered that all the Male Labouring Tithable Persons Convenient thereto Assist in Clearing and keeping the same in Repair.

7 March 1747 O. S., Page 393
Ordered that a Road be laid off and Cleard the best and Most Convenient way from the Ridge Road that leads from Rutledges ford to the Bridge over Little Ronoke, and Richard Hill is apointed Surveyor thereof, And 'tis Ordered that all the Male Labouring Tithable Persons Convenient thereto Assist in Clearing the same.

7 March 1747 O. S., Page 394
Ordered that Lewis Delony Gent hire some Person to View the best and most Convenient Place for a Bridge over Maherrin River between the Fork and Mizes Ford and make Report thereof with the Charge of such hire here to the Next Court.

7 March 1747 O. S., Page 395
Ordered that a Bridge be built over Little Ronoke Where the Road leading to Youngs Mill crosses the same.

7 March 1747 O. S., Page 395
Ordered that a Road be laid off and Cleard the Best and most Convenient way from Howards Road crossing Miles's Creek, to Mizes Ford, and John Watson is appoint'd Surveyor thereof, And 'tis Ordered that all the Male Labouring Tithable Persons Convenient thereto Assist in Clearing the same.

7 March 1747 O. S., Page 396
Drury Melone is appointed Surveyor of the Road called Howards Road from the County up to the great Meadow, And it is Ordered that all the Male Labouring Tithable Persons Convenient thereto Assist in Clearing and keeping the same in Repair.

7 March 1747 O. S., Page 396
Valentine Brown is appointed Surveyor of the Road leading from the North River to the South River, and 'tis Ordered that all the Male Labouring Tithable Persons Convenient thereto Assist in Clearing and keeping the same in Repair.

7 March 1747 O. S., Page 396
John Cox is appointed Surveyor of the Road leading from the South River to this Courthouse & 'tis Ordered that all the Male Labouring Tithable Persons Convenient thereto Assist in Clearing the same

4 April 1748 O. S., Page 397
Upon the Petition of Hampton Wade & others, It is Order'd that a Road be laid off and Clear'd the best and most Convenient way from Nottoway Road, tending across F***king Creek, to Maherrin Road leading to this Courthouse, and Thomas Wynne is Appointed Surveyor thereof, And 'tis Ordered that all the Male Labouring Tithable Persons Convenient thereto Assist in Clearing the same.

4 April 1748 O. S., Page 398
On the Petition of George Currie Gent Leave is given him to Keep a Ferry over Great Ronoke from Munfords Quarter to the Ochenechee, Provided he enter into Bond with Security According to Law, in the Clerks office and he is allow'd for a Man and Horse Seven Pence half Penny, And on the Prayer of the said George Currie leave is given him to Clear a Road from his Ferry into Cooks Courthouse Road at his own Cost and Charge.

4 April 1748 O. S., Page 400
Ordered that Lewis Delony Gent, Thomas Easland John Williams and Joseph Blanks do View a place Suitable and Convenient for a Bridge over Maherrin and make Report thereof to the next Court.

4 April 1748 O. S., Page 401
Mathew Talbot, John Caldwell, and William Caldwell Gent are appointed to agree with Workmen to Build a Bridge heretofore Ordered to be Built over Little Ronoke, And that they View the best and most Convenient place between Youngs Mill and Martins ford at or Nigh the great Lick where the Road crosses the said River whereof John Young was Surveyor and make Report thereof here to the Next Court.

2 May 1748 O. S., Page 424
Lewis Delony & Thomas Eastland two of the Persons appoint'd to View the best and most Convenient place for a Bridge over Maherrin River, this day Returned their Report which is Ordered to be Recorded.

2 May 1748 O. S., Page 424
John Williams and Joseph Blanks two of the Persons appointed to View the best and most Convenient Place for a Bridge over Maherrin River, this day Return'd their Report which is Ordered to be Recorded.

2 May 1748 O. S., Page 426
Ordered that Lewis Delony and Clement Read Gent do agree with Workmen to Build a Bridge over Maherrin River at the Place Reported by the said Lewis Delony & Thomas Eastland upon such Terms and in such manner as they shall think fit, And make Report thereof here to the Next Court.

6 June 1748 O. S., Page 9
The Proposals and Orders for Building a Bridge over Maherrin River were Exhibited and Ordered to be Recorded.

6 June 1748 O. S., Page 9
Hugh Lawson Gent and Samuel Wilson two of the Persons appointed by the last Court to view a place on the North River for a Bridge, this Day return'd their Report thereon which is Ordered to be Recorded.

6 June 1748 O. S., Page 12
Ordered that a Road be laid off and Clear'd the best and most Convenient way from John Phelps's Mill crossing Seneca a little below the fork, thence a little below the three forks of Falling River and from thence Runing a little below the fforks of Cubb Creek to Intersect Robert Bakers Road, And the said John Phelps is appointed Surveyor thereof from his Mill to the fforks of Seneca, And it is Ordered that all the Male Labouring Tithable Persons Convenient thereto, Assist in Clearing and keeping the same in Repair.

6 June 1748 O. S., Page 13
William Thomas is appointed Surveyor of the New Road leading from the fforks of Seneca to ffalling River, And 'tis Ordered that all the Male Labouring Tithable Persons Convenient thereto, Assist in Clearing & keeping the same in Repair According to Law.

6 June 1748 O. S., Page 13
John Read is appointed Surveyor of the New Road leaving falling River to where it Intersects Robert Bakers Road and it is Ordered that all the Male Labouring Tithable Persons Convenient thereto, Assist in Clearing and keeping the same in Repair According to Law.

6 June 1748 O. S., Page 13
Ordered that a Road be laid off and Clear'd the best nearest and most Convenient way from or near John Stewarts on Cubb Creek to the fforks of Falling River, And William Rogers is appointed Surveyor thereof And it is Ordered that all the Male Labouring Tithable Persons Convenient thereto forthwith Assist in Clearing and keeping the same in Repair According to Law.

3 October 1748 O. S., Page 80
David Gwin is appointed Surveyor of the Road Whereof John Young decas'd was late Surveyor, And 'tis Ordered that all the Male Labouring tithable persons Convenient thereto Assist in Clearing and Keeping the same in Repair According to Law.

5 February 1748 O. S., Page 102
Ordered that a Road be laid off and Cleared the nearest and best way from the Ridge road that leads from Rutledges foard on Appomattox River to the Bridge over Little Ronoke, and Richard Hill is appointed Surveyor thereof and it is ordered he with the Male Labouring Tithable persons following, to wit, Owen Sullivant, Daniel Sullivant Jacob Stober, Ruben Roy, Cap[t] John Ruffins Tithables at his Quarter, John Mount, John Mullins Stephen Collins & his, Thomas Rice, Charles Harris and those of Nanny Dabbs forthwith Clear and keep the same in Repair According to Law.

5 February 1748 O. S., Page 102
David Gwinn is appointed surveyor of the Road leading from Gwinns Bridge to Young's Mill and it is Ordered that he with all the Male Labouring Tithable Person Convenient thereto Assist in Clearing and keeping the same in Repair According to Law.

5 February 1748 O. S., Page 106
Ordered that James Hunt & Charles Talbot View the Several Roads leading over Cubb Creek, and report to the Next Court Which they think is the most Convenient to Build a Bridge on over the said Creek.

5 February 1748 O. S., Page 107
Thomas Bouldin & Clement Read Gent are appointed to agree with Workmen to Build the Bridges over Little Ronoke and Wards fork upon such Terms and in such Manner and form as they shall think fit.

6 February 1748 O. S., Page 127
Ordered that David Stokes, Lyddall & Hugh Lawson Gent do agree with Workmen to Build a Bridge over F***ing creek in this County at such place upon Terms and in such manner and form as to them shall seem fit and Convenient.

5 June 1749 O. S., Page 151
Henry May is appointed Surveyor of the Road leading from the Mouth of Little Ronoke to the Church along Randolphs Road, And it is Ordered that he with the Male Labouring Tithable Persons under Robert Williams and those of M^r. James Cocke, forthwith Clear and keep the same in Repair According to Law.

5 June 1749 O. S., Page 154
George Foster is appointed Surveyor of the Road whereof Richard Womack was late Surveyor, And it is Ordered that he with all the male Labouring Tithable Persons Convenient thereto, forthwith Clear and keep the same in Repair According to Law.

5 June 1749 O. S., Page 154
Daniel Firth is appointed Surveyor of the Randolphs Road from Thomas Worthys to the Mossing foard, And it is Ordered that the said Firth with all the Male Labouring Tithable Persons under the said Thomas Worthy, Thomas Handcock and his, William Brumfield, John Towns and

his, John Gwinn and his, Charles Talbot and his, Richard Jones, Joseph Jones, Michael McDaniel, and Henry Howard, forthwith Clear and keep the same in Repair According to law.

5 June 1749, O. S., Page 154
Ordered that a Road be laid off and cleared the best and most Convenient Way from Stanton River to the Mayo Settlement at the Wart Mountains And Joseph Mayse is appointed Surveyor of that Part thereof as leads from Stanton River to Allens Creek, And it is Ordered that the said Mayse with all the Male Labouring Tithable Persons Convenient thereto, forthwith lay open Clear and keep the same in Repair According to Law.

5 June 1749 O. S., Page 155
Richard Parsons is appointed Surveyor of Part of the New Road leading from Stanton River to the Mayo Settlement at the Wart Mountains, to wit, from Allens Creek to Banister River, And tis Ordered that the said Parsons with all the Male Labouring Tithable persons Convenient to the said Road forthwith mark off and lay open the best and most Convenient Way and Clear and keep the same in Repair According to Law.

5 June 1749 O. S., Page 155
Elisha Walling is appointed Surveyor of Part of the New Road leading from Stanton River to the Mayo Settlement at the Wart Mountains, to wit, from Banister River to Smiths River, And 'tis Ordered that the said Walling with all the Male Labouring Tithable persons Convenient to the said Road forthwith mark off and lay open the best and most
Convenient way, and Clear and keep the same in Repair According to Law.

5 June 1749 O. S., Page 155
Joseph Cloud is appointed Surveyor of Part of the New Road leading from Stanton River to the Mayo Settlement of the Wart Mountains, to wit, from Smiths River to the said Settlement, And it is Ordered that the Said Cloud with all the Male Labouring Tithable persons convenient to the said Road forthwith mark off and lay open the best and most Convenient way, and Clear and keep the same in Repair According to Law.

5 June 1749 O. S., Page 156
Ordered that a Road be laid off and Clear'd the best and most Convenient way from Youngs Mill Path to Little Ronoke Bridge at Coles Road, And Samuel Duval is appointed Surveyor thereof, And it is Ordered that the said Duval with all the Male Labouring Tithable Persons Convenient thereto, to wit, Samuel Johnson, William Johnson, Stephen Collins, and his, John Pratt, Clement Reads tithables, Thomas Vernon, John Austin, John Atkins, John Austin Jur, Richard Wards Tithables, and all others Convenient thereto forthwith Clear and keep the same in Repair According to Law.

5 June 1749 O. S., Page 156
John Mead is appointed Surveyor of the Road leading from the Fish Dam on Otter River to the Meadows, And it is Ordered that the said Mead with all the Male Labouring Tithable Persons Convenient to the Said Road on the North side of Stanton River forthwith Clear and keep the same in Repair According to Law.

5 June 1749 O. S., Page 156
William Morgan is appointed Surveyor of the Road leading from the Meadows to Maggotty Creek And it is Ordered that the said Morgan with all the Male Labouring Tithable Persons Convenient to the said Road on the South side of Stanton River and upon Maggotty Creek as high as John Millers, forthwith Clear and keep the same in Repair According to Law.

5 June 1749 O. S., Page 157
Joseph Rentfro is appointed Surveyor of the Road leading from the Magotty Creek to the Burying Place to the End of the Road, And it is Ordered that the said Rentfro with all the Male Labouring Tithable Persons Conveneint to the said Road forthwith Clear and Keep the same in Repair According to Law.

4 July 1749 O. S., Page 186
William Rodgers on Falling River, is Appointed Surveyor of the Road leading from Turnip Creek to Falling River in the Room of James Mcdade, and it is Ordered that the said Rodgers, together with the hands that Assisted the said M^{c}dade, do forthwith Clear & keep in Repair the said Road according to Law.

4 July 1749 O. S., Page 187
Richard Hix is Appointed Surveyor of the Road leading from Thomas Bouldins to Jones's new Road, and it is Ordered that he together with all the Male Labouring Tithables Convenient thereto, do forthwith lay open Clear & keep the same in Repair According to Law.

4 July 1749 O. S., Page 188
Reese Presse Senr. is Appointed Surveyor of the Road leading from Otter River to the Popler Spring, whereof John Mead was late Surveyor, and it is Ordered that he together with the hands that Assisted the said Mead thereon, do forthwith Clear and keep the Same in Repair According to Law.

4 July 1749 O. S., Page 188
Daltons Pet ab. Roads Granted

4 July 1749 O. S., Page 190
On the Motion of John Boyd, Licence is Granted him to keep a Ferry over Dann River in this County, and the Court do adjudge that he take four Pence Ferriage for a man and four pence for a Horse; and for Wheal Carriages for each wheal, the same as for a Horse

6 July 1749 O. S., Page 202
Ordered that Henry Embrys, John Bacons, Edward Caldwellls and Joseph Minors hands Assist in clearing and keeping in Repair the Road leading from Reedy Creek to Dyors Pine.

3 October 1749 O. S., Page 207
On the Petition of Richard Randolph Licence is granted him to keep a Ferry over Roanoak River at the Mouth of Little Roanoak River in this County, and it is Considered that he take Six pence Feriage for a Man & Six pence for a Horse and for Wheel Carriages According to Law.

3 October 1749 O. S., Page 208
David Gwin is Appointed Surveyor of the Road leading from Randolphs Road to Youngs Mill, and it Ordered that he together with all the male Labouring Tithables Convenient thereto forthwith Clear and keep the same in Repair According to Law.

3 October 1749 O. S., Page 209
Ordered that John Caldwell and Thomas Dorrity do Divide and Proportion the hands to assist the following Persons in Clearing and keeping in Repair the Roads whereof they are Surveyors, (as the said John and Thomas shall think fit) to wit, Richard Dudgeon, David Caldwell, William Rodgers, George Abney, and Israel Pickens.

3 October 1749 O. S., Page 209
Ordered that a Bridge be Built over Nottoway River between Abraham Cooks and William Crosses, and it is Ordered that Hugh Lawson Gent. do make aplication to the County Court of Amelia to Joyn in Building Such Bridge.

3 October 1749 O. S., Page 209
On the motion of Abraham Cook licence is Granted him to keep an Ordinary at the Forks of the Road near his Plantation in the County Provided he enter into Bond with Security for that Purpose in the Clerks Office some time between this and the next Court.

3 October 1749 O. S., Page 210
Philip Jones is Appointed Surveyor of the Road leading from the Fork of Falling River Road to David Jones's Road, and it is Ordered that he together with all the Male Labouring Tithables Convenient thereto do forthwith Clear and keep the same in Repair According to Law.

3 October 1749 O. S., Page 210
Ordered that a Bridge be Built over Banister River at the Cow Ford in this County, and it is Ordered that Abraham Martin and Cornelius Cargill Gent do let the same to undertakers on such Terms and Conditions and for what Considerations they shall think fit.

3 October 1749 O. S., Page 210
Ordered that a Bridge be Built over Ledbetter Creek in this County and it is Ordered that David Stokes and Lyddal Bacon Gent[l] do let the same the same to undertakers upon such Terms and Conditions and for such Consideration as they shall think fit.

3 October 1749 O. S., Page 210
John Binum is appointed Surveyor of the Road leading from little Cheary Stone to Allen Creek and it is Ordered that he together with all the Male Labouring Tithables Convenient thereto do forthwith Clear and keep the same in Repair According to Law.

3 October 1749 O. S., Page 210
Jeremiah Hatcher is Appointed Surveyor of the Road leading from the Bares Element Creek to the Middle fork of Maherrin River, and it is Ordered that he together with the all the Male Labouring Tithables Convenient thereto do forthwith Clear and keep the same in Repair According to Law.

3 October 1749 O. S., Page 211
John Twitty is Appointed Surveyor of the Road leading from the Middle Fork of Maherrin River to this Court House, and it is Ordered that he together with all the Male Labouring Tithables Convenient thereto do forthwith Clear and keep the same in Repair According to Law.

4 October 1749 O. S., Page 221
Grand Jury Presentments
... a Presentment against the Surveyors of the Road from Randolphs Road by Samuel Perrins to the Court House for not keeping it in Repair ...

13 November 1749 O. S., Page 248
County Levy
To Abraham Martin for building a Bridge Over Little Roanoak River
    L 7.3.0

To Henry Isbell for building a Bridge Over Little Roanoak River on Coles Road when the shall be Received_____ 5.15.0

* * *

To David Caldwell for Building a Bridge over Cubb Creek ....
> 20.0.0

To David Caldwell for Timber for d° ...
> 2.0.0

To Richard Stokes for a Bridge over Maherrin River ...
> 6.0.0

2 January 1749 O. S., Page 253
William Hill Gent[l] is Appointed Surveyor of the Road Leading from Capt Mitchells Landing to this Court House, and it is Ordered that he together with all the Male Labouring Tithable Persons Convenient thereto, do forthwith Clear and keep the same in Repair According to Law.

2 January 1749 O. S., Page 253
Joseph Blanks is Appointed Surveyor of the Road leading from Rays Bridge to Flat Rock Road, and it is Ordered that he together with John Edloes Hands, Joseph Blanks hands, Henry Tally's, Francis Ray's, James Dawes, William Allen, William Saffold, Henry Soans, John Forrest, Edward Jackson, Israel Brown, Edward Whit & Cap[t] William Bookers hands do forthwith Clear and keep the same in Repair According to Law.

3 January 1749 O. S., Page 258
On the Petition of George Walton and others, It is Ordered that a Road be lay'd open and Cleared from Randolphs Road a little above George Moores the nearest and best way to the Mouth of Finny Wood, and Thomas Nance is Appointed Surveyor thereof from Randolphs Road to Robertsons Fork of Meherrin River, and it is Ordered that he together with Peter Rowlet, Robert Haislop, John Nance Ju[r]. Richard Nance Jun[r]. Richard Thompson John Willingham Jun[r]. Thomas Crenshaw James Breedlove and George Waltons Male Labouring Tithables, do forthwith lay open, Clear and keep the same in Repair According to Law, and Joseph Williams is Appointed Surveyor thereof from the Robertsons Fork of Meherrin River to the Mouth of Finny Wood, and it is Ordered that he together with the Male Labouring Tithables at Johnsons Quarter, William Dobbyns's and Silvanus Stokes's Male Labouring Tithable Persons do forthwith lay open Clear and keep the same in Repair According to Law.

4 January 1749 O. S., Page 273
William Williams is Appointed Surveyor of the Road leading from Miles's Creek to Delonys Mill whereof Lewis Delony was late Surveyor, and it is Ordered that he together with the hands that Assisted the said Delony thereon do forthwith Clear and keep the same in Repair According to Law.

4 January 1749 O. S., Page 273
William Williams is Appointed Surveyor of the Road leading from Howards Road to Rays Bridge, it is Ordered that he together with all the Male Labouring Tithable Persons Convenient thereto do forthwith lay open the same --

4 January 1749 O. S., Page 273
Francis Ellidge is Appointed Surveyor of the Road leading from Delonys Mill to Allens Creek, and it is Ordered that he together will all the Male Labouring Tithable Persons convenient thereto do forthwith Clear and keep the same in Repair according to Law.

NOTE: At the end of the minutes of the January Court it said that the Court will be adjourned until the second Tuesday in April next but on the following page the notes resume for the July Court rather than the April Court. Therefore, there may be a significant number of missing road orders.

3 July 1750 O. S., Page 277
Thomas Nash is Appointed Surveyor of the Road leading from the Middle Bridge over Little Roanoak River to the upper Bridge, and it is Ordered that he together with all the Male Labouring Tithable Persons Convenient thereto do forthwith Clear and keep the same in Repair According to Law.

3 July 1750 O. S., Page 288
Thomas Staples is Appointed Surveyor of the Road whereof John Nance was late Surveyor, and it is Ordered that he together with the hands that Assisted the said Nance thereon do forthwith Clear and keep the Same in Repair According to Law.

3 July 1750 O. S., Page 293
Ordered that a Road be laid open and Cleared the most Convenientest Way from the Main Road near Hueys Cart path towards Hawkins's fferry as far as the Country line and Benjamin Harrison is Appointed Surveyor thereof, and it is Ordered that he together with William Douglass and Robert Brooks & his Male Labouring Tithables do forthwith lay open Clear and keep the same in Repair According to Law.

4 July 1750 O. S., Page 295
John Denny is Appointed Surveyor of the Road leading from the fork of Sincker to the fork of Falling River whereof William Thomas was late Surveyor, and it is Ordered that he together with the hands that Assisted the said Thomas thereon do forthwith Clear and keep the same in Repair According to Law.

4 July 1750 O. S., Page 304
On the Motion of Joseph Mayse Licence is Granted him to Keep a Ferry in this County on Roanoake River on his Bond & security in the Clerks Office for that Purpose, and the Court do adjudge that he shall Take four pence Ferriage for a Man & eight pence for a Man & Horse & No more --

5 July 1750 O. S., Page 316
William Been is appointed Surveyor of that part of the Road Leading from his House to the Head of Sandy River & it is Ordered that all the Hands Convenient thereto Assist the said William in Clearing & Keeping the same in Repair according to Law.

5 July 1750 O. S., Page 317
Tidance Laine is appointed Surveyor of that Part of the Road Leading from the Head of Sandy River to the Lower falls of Banister River. & it is Ordered that all the Hands Convenient thereto Assist the said Tidance in Clearing & Keeping the same in Repair according to Law.

5 July 1750 O. S., Page 317
William Russell is appointed surveyor of that Part of the Road Leading from the Lower falls of Bainster River to Fuqua's Ford & it is Ordered that all the Hands Convenient Thereto do Assist the said William in Clearing and Keeping the same in Repair According to Law.

2 October 1750 O. S., Page 346
On the Motion of Welcom William Hodges & others, Ordered that a Road Laid Open & Cleared the Nearest best & Most Convenient way from the Mouth of Elk horn Creek to Mayses ford on Stanton River, & it is Ordered that William Garrot Together with all the Hands Convenient do forthwith Clear & Keep the sam in Repair according to Law.

2 October 1750 O. S., Page 347
The Petition of Robert Brookhouse & others for a ferry, for Reasons appearing to the Court it is Rejected --

2 October 1750 O. S., Page 347
John Logan is appointed Surveyor of the Road whereof Charles Talbot was Late Surveyor and it is Ordered that he Together with all the Assistance that Assisted the said Charles Talbot do forthwith Clear & Keep the same in Repair according to Law.

2 October 1750 O. S., Page 347
Ordered that William Williams build a Bridge over Merherin River at Mices ford in Leiu of the Bridge he built thereon which is Lately Carried away by freshes, & that he Compleat the same by the tenth day of June Next --

2 October 1750 O. S., Page 348
On the Petition of James Hunt Gentlemen & Others It is Ordered that a Bridge be Built over Turnip Creek where Coles Road Crosses the said Creek in this County, & it is Ordered that John Caldwell William Caldwell, & James Hunt Gentlement treat with workmen to build the same upon such terms and such Manner, & for what Consideration they shall think Proper.

3 October 1750 O. S., Page 358
Andrew Martin is appointed Surveyor of the Road Whereof Israel Pickings was late Surveyor, and it is Ordered that he Together with the Hands that Assisted the said Pickens on the said Road do forth with Clear & Keep the same in Repair According to Law.

3 October 1750 O. S., Page 358
Hutchins Burton is appointed Surveyor of the Road Whereof Field Jefferson Gentl. was late Surveyor, and it is Ordered that he Together with the Hands that Assisted the said Jefferson thereon do forth with Clear & Keep the same in Repair According to Law.

3 October 1750 O. S., Page 359
Michael Johnson is appointed Surveyor of the Road Whereof Joseph Williams was late Surveyor and it is Ordered that he Together with with the Hands that Assisted the said Williams thereon do forthwith Clear & Keep the same in Repair According to Law.

3 October 1750 O. S., Page 359
Ordered that a Bridge be Built over the fork of Meherrin River in this County, and it is Ordered that Henry Embry, Richard Witton, & John Twitty Gentlemen or any two or more of them Treat With Workment to Build the same upon such Terms in Such Manner & for what Consideration they shall think Proper.

3 October 1750 O. S., Page 367
Thomas Worthey is appointed Surveyor of the Road Whereof Daniel Firth Deceased, was late Surv^r. and it is Ordered that he Together with the Hands that Assisted the said Firth thereon do forthwith Clear & Keep the same in Repair According to Law.

2 April 1751 O. S., Page 377
Julius Nichols is appointed Surveyor of the Road Leading from Dockerys to his Ferry on Stanton River, and it is Ordered that he Together with the Following Assisstance (to Wit) William Humprey, Humphrey Hughey, John Grissel, John Cole, Gabriel Hardin, William Hardin, Thomas Weeks, Michael Weeks, David Dodd, Amos Timms, William Nichols, John Wilson, Reuben Morgan & Philip Morgan, with their Male Labouring Tithable Persons Do forthwith Clear and Keep the Same in Repair According to Law.

2 April 1751 O. S., Page 381
Ordered that a Road be laid off & Cleared the best and most Convenient way from the Place where James Caldwells Path turns out of the Road to Wards fork Bridge, and Andrew Martin is appointed Surveyor thereof, and it is Ordered that he Together with all the Male Labouring Tithables Convenient thereto do forthwith Law open Clear & Keep the same in Repair according to Law.

2 April 1751 O. S., Page 382
On the Petition of Francis Elledge & others Leave is Granted them to Lay open & Clear a Road to Strike out of Mitchells Road, about Half a Mile Below Henry Sages the best & Most Convenient way into a Road Made Use of by Richard Wilton Gentleman Crossing Meherrin River below the Middle fork and thence into Twittys Road, & that the Petitioners Clear the same & Not to be Discharged of other Roads

2 April 1751 O. S., Page 383
Thomas Rodgers is appointed Surveyor of the Road Whereof William Rodgers was Late Surveyor & it is Ordered that he together with the Assistance that assisted the said William on the said Road do forthwith Clear and Keep the same in Repair according to Law.

2 April 1751 O. S., Page 384
William Caldwell & Thomas Daugharty are appointed to devide the Hands on the Roads Mentioned in a former Ordered of this Court

2 April 1751 O. S., Page 384
On the Petition of Thomas Anderson & Others It is Ordered that James Mitchell Hugh Lawson Field Jefferson & Lyddal Bacon Gentlemen (being first Sworn &c) do diligently Veiw & Examine a way where a Road is lately Petitioned for, and Twitty's Road Crossing the Middle fork of Meherrin River & report hereto the Next Court which is the Best Nearest & Most Convenient way

2 April 1751 O. S., Page 384
Ordered that a Road be Laid open & Cleared from the Mouth of Blue stone Creek the Best & Most Convenient way into the Road Leading from this Courthouse to Twitty's Ordinary.

2 April 1751 O. S., Page 385
On the Motion of James McDaniel Ordered that a Road be Laid open & Cleared the best and Most Convenient way from the Cherry Stone Road over Meherrin River at the Mouth of Little Beaver Pond to Crooked Creek Bridge, it is Ordered that Jeffrey Russel, John Andrews, William Allen, Edward Whitt & James McDaniel & their Male Labouring Tithables do forthwith Lay open Clear & Keep the same in Repair According to Law.

2 April 1751 O. S., Page 385
John Wilborne is appointed Surveyor of the Road leading from the Mouth of Blue Stone Creek to Twitty's Ordinary, and it is Ordered that he Together with the follow^g. Assisstance (to wit) John Clarke, William Lidderdale, John Cox, John Thompson, and all William Byrds Hands below the Mouth of Blue Stoke Creek, do forthwith Clear & Keep the same in Repair according to Law.

2 April 1751 O. S., Page 385
John Cargill is appointed Surveyor of the Road leading from Stanton River to this Courthouse, & it is Ordered that he Together with all the Hands on Dan River in the Fork as High as Andrew Wades and on Stanton River in the fork as High as Tandy Walkers & all the Hands on the Lower side of the said River Convenient do forthwith Clear & Keep the same in Repair According to Law.

2 April 1751 O. S., Page 389
Richard Wilton & Hugh Lawson Gentlemen are Appointed & Desired to Meet Charles Irby & Abraham Cooke the Gentlemen appointed by Amelia Court to Let the Building of a Bridge over Notaway River that Divides this & Amelia County in Order to Let the Building the said Bridge to Undertakers on such terms & Conditions and for such Consideration they the said Richard Wilton Hugh Lawson Charles Irby & Abraham Cooke Gentlemen shall think fitt.

3 April 1751 O. S., Page 390
On the Petition of Thomas Brandon & Others, it is Ordered that a Road be laid open & Cleared the best & Most Convenient way from the County line Creek to fishing Creek & Thomas Wise is Appointed Surveyor thereof, and it is Ordered that he together with all the Hands Convenient do forthwith Clear & Keep the same in Repair according to Law.

3 April 1751 O. S., Page 390
On the Petition of Thomas Brandon & others Ordered that a Road be laid open & Cleared the Best & Most Convenient way from fishing Creek to John Boyds Ferry & James Irvine is appointed Surveyor thereof, & it is Ordered that he Together with all the Hands Convenient do forthwith Clear & Keep the same in Repair according to Law.

3 April 1751 O. S., Page 391
John Saunders is appointed Surveyor of the Road Whereof Joseph Mayse was late Surveyor, and it is Ordered that he Together with the Hands that Assisted the said Joseph on the said Road, do forth Clear & Keep the same in Repair according to Law.

3 April 1751 O. S., Page 391
On the Petition of Robert Wade Licence is Granted him to Keep a Ferry over Standton River in this County from his Land to the Land of Cornellius Cargill Gentleman On his Giveing Bond & Security in the Clerks Office of this County some time Between this & the Next Court according to Law, and the said Wade doth Undertake to make the said Ferry free to all the inditants of This County to Pass & Repass to & from this Court and the Churches in the said County & at all other Adjudje times & to other People the Court do adjudje that he shall Take four pence ferriage for a Man & eight Pence for a Man & Horse and no More and the Court also agrees to Allow the said Wade one Thousand Pounds of Tobacco for Ferriage of the Inhabitants of this County to Court & Church & c^a. for the ensuing Year.

3 April 1751 O. S., Page 393
Thomas Eastland is appointed Surveyor of the Road Leading from Allens Creek to Butchers Creek Whereof William Howard was late Surveyor, and it is Ordered that he Together with Assisstance that Assisted the said William Howard thereon do forth with Clear & Keep the same in Repair According to Law

3 April 1751 O. S., Page 393
William Caldwell Gentlemen is Appointed & Desired to agree With John Austin to Repair the Bridge over Wards fork upon such conditions and for what Consideration as he shall think Proper

3 April 1751 O. S., Page 394
Ordered that a Bridge be Built over Allens Creek at the old Bridge In this County and Field Jefferson & James Mitchell Gentlemen are appointed & Desired to Treat with Workmen to build the same in Such Terms & Conditions in Such Manner and for what Consideration they shall think Proper

3 April 1751 O. S., Page 394
On the Motion of William Marable & others it is Ordered that a Bridge be Built over Meherrin River above Scotts where the Road Crosses the said River, & it is Ordered that Henry Embry and Lyddal Bacon Gentlemen treat with Workmen to Build the same on such Terms in such manner and for What Consideration they shall think Proper

3 April 1751 O. S., Page 394
On the Petition of William Caldwell Gentleman & others. It is Ordered that a Road be laid Open & Cleared the Best & Most Convenient Way from the firsh dam on Otter River to or Near the Mouth of Goose Creek and William Verdiman is appointed Surveyor thereof, & it is Ordered that he Together with all the Male Labouring Tithables on Stanton River from the Pocket to Where James Timker Formerly Lived do forthwith Clear & Keep the same in Repair According to Law. & that they be Exempted from Working on the old Road.

3 April 1751 O. S., Page 395
Ordered that a Road be Laid Open & Cleared the Best & Convenient way from the Ford On Otter River Near the Mouth of Elk Creek down by William Callaway's Mill to the Ridge that Divide this County from Albemarle where Rusts Path Crosses the said Ridge. William Callaway Gentl is appointed Surveyor thereof and it is Ordered that he Together with all the Hands Convenient thereto do forthwith Clear & Keep the same in Repair According to Law

3 April 1751 O. S., Page 395
Ordered that a Road be laid open & Cleared the Best & Most Convenient Way from the Foot of Johnsons Mountain (Lying on the south side of Otter River) into Callaways Road & Richard Callaway is Appointed Surveyor thereof & it is Ordered that he Together with all the Hands Convenient do forthwith Clear & Keep the same in Repair according to Law

4 April 1751 O. S., Page 411
William Marable is appointed Surveyor of the Road Leading from his Ordinary to this Courthouse & it is Ordered that he together with all the Hands Convenient thereto do forth Clear & Keep the same in Repair according to Law

4 March 1751 O. S., Page 411
Mackness Goode is appointed Surveyor of a Road leading from Marables Ordinary to Kings Road & it is Ordered that he Together with all the Hands Convenient do forthwith Clear & Keep the same in Repair according to Law

4 April 1751 O. S., Page 411
Henry Blagrave is appointed Surveyor of the Road leading from a North fork of Meherrin River to the Middle Fork & it is Ordered that he together with all the Hands Convenient thereto do forthwith Clear & Keep the same in Repair according to Law

4 April 1751 O. S., Page 419
Ordered that a Bridge be Built over blue Stone Creek where the Road leading from Wades Ferry Crosses the said Creek & Abraham Martin & Cornellius Cargill Gent[l]. are appointed and Desired to let the same to Undertakers in such Manner upon such Terms & Condition & for what Consideration the shall think proper

4 April 1751 O. S., Page 420
Ordered that a Bridge be build over the Great Creek where the Treading Path leading to Nichols's Ferry Crosses the said Creek & Field Jefferson & James Mitchell Gentl are Desired & appointed to lett the same to Undertakers On Such terms & Conditions in such Manner & for what Consideration as they shall think fit.

5 April 1751 O. S., Page 421
On the Motion of Julius Nichols leave is granted him to Keep a ferry at the Place where Hickman formerly Kept a Ferry & the Court do adjudge that he have the same allowances as were allow'd the said Hickman

5 April 1751 O. S., Page 437
Ordered that a Road be laid open & Cleared the Best & Most Convenient way from John Cargills to Cox's Road & McNesse Goode is appointed overseer thereof, & it is Ordered that he together with all the Hands Convenient thereto do forth with Clear & Keep the same in Repair according to Law.

2 October 1751 O. S., Page 461
On the Motion of John McNess he is Exempted from being Overseer of the Road Leading from Coxes Road to Cargills Road whereof he was Late Surveyor

2 October 1751 O. S., Page 462
On the Petition of William Caldwell & others, for a Road to be Laid off and Clear'd the best and Convenintest way from Cubb Creek Bridge to Bouldins Road, And it is ordered that Robert Woods late Surveyor of the Road Leading from Cubb Creek to Little Roanoake and Captain Bouldins together with the usual assistance that assisted him in Clearing and keeping the same in Repair do forth with assist the s^d. Robert in Clearing and laying open the said Road

2 October 1751 O. S., Page 464
Order'd that John Knight open and Clear a Road from the Head of Montaine Creek to Nottaway River the best and Most Convenientest way, by John Knights and it is ordered that the said Knight together with the following Assistance towit, Tscharner Degrantenreidt & his Hands, John Knights and his, Mrs Fishers Hands Kemp & his Hands, & Joseph Minor & his hands & Coris Christopher with their Male Labouring Tithables Persons do forth with Clear & keep the same in Repair According to Law.

2 October 1751 O. S., Page 464
William Williams is appointed Surveyor of the Road leading from Smiths River to Leatherwood and it is ordered that he together with the Male Labouring Tithable Person Convenient thereto do forth with Clear and keep the same in Repair according to Law.

2 October 1751 O. S., Page 464
Merry Webb is appointed Surveyor of the Road leading from Leatherwood to the north fork of Sandy River, And it is ordered that he together with the Male Labouring Tithable Persons Convenient thereto do forth with assist the said Webb in Clearing and keeping the same in Repair according to Law.

2 October 1751 O. S., Page 464
Ruben Rey is appointed Surveyor of the Road leading from the north fork of Sandy River to Bear Skin. And it is ordered that he together with the Male Labouring Tithable Persons Convenient thereto do forth with assist the said Ruben in Clearing and keeping the same in Repair According to Law.

2 October 1751 O. S., Page 465
Richard Parsons is appointed Surveyor of the Road leading from the Bear Skin to little Cherry Stone and it is ordered that he together with the Male Labouring Tithable Persons Convenient thereto do forthwith Assist the said Richard in Clearing and keeping the same in Repair According to Law

2 October 1751 O. S., Page 466
James Amos is appointed Surveyor of the Road leading from Stokes's Bridge to the Road that Crosses William Crosses Bridge and it is ordered that he together with the Male Labouring Tithable Persons Convenient thereto do forthwith assist the s^d Amos in turn the said Road from the South Side of Samuel Jones Plantation to the North side of Robert Liverits from Crosses Bridge to the fork of the Road above Hound's Creek Race Paths and keep the same in Repair According to Law

2 October 1751 O. S., Page 468
James Cocke is appointed Surveyor of the Road Whereof John Ashworth was late Surveyor, And it is ordered that he together with the Assistance that Assisted the said Ashworth on the said Road, do forth with Clear and keep the same in Repair According to Law

2 October 1751 O. S., Page 469
Order'd that Abraham Martin Gent. together with the following Assistance, to wit, Robert Williams and his Hands Nehemiah Frank, John Sansum, and Abram Martins hands do forthwith open and Clear a Road the best and Most Convenients Way from the said Martins to Kings Road, And

it is ordered that they assist him in keep the same in Repair According to Law.

2 October 1751 O. S., Page 469
Order'd that Thomas Bouldin John Gwinn and Charles Talbot or any two of them they being first Sworn & c do run & Examine the way from the Mossing Ford into M^r. Martins Road and Report to next court the Conveniency or inconveniency thereof

2 October 1751 O. S., Page 470
Ordered that a Road be laid off and Cleared the Best and most Convenientest way from Nicholss Ferry to the County Line to Meeat the Road leading from Granvill Courthouse to William Eatons, and John King is appointed Surveyor thereof, And it is ordered that he together with the following Assistance to wit William King, John Simmons, and all the Male Labouring Tithable Persons at the Plantations of Baxter and Edward Davis's do forthwith assist the said John King in Clearing and keeping the same in Repair According to Law

2 October 1751 O. S., Page 470
On the Petition of Tscharner Degrantenreidt Gent. for a Road to be laid off and Clear'd the best and most Convenient Way from Willingham's Road to Wynn's Road, It is ordered that Tscharner Degrantenreidt, George Walton and Joseph Billups (or any two of them being first sworn &c) do view and Examine the way where such Road is Petitioned for, and Report to the next the Conveniency or Inconveniency thereof

2 October 1751 O. S., Page 471
On the Petition of Edward Mobberly and others for a Road to laid of and Cleared the best and Most Convenientest way from the Church on Otter River to the said River to Clement Mobblerlies from thence to the South fork of little Otter River and from thence to Goose Creek near John Harvies, It is ordered that Edward Mobberley & John Harvey (being first Sworn) do view & Examine the way where such Road is Petitioned for, and Report to the next court the Conveniency and Inconveniency thereof.

2 October 1751 O. S., Page 474
Peter Hudson is appointed Surveyor of the Road leading from Graves ford to John Ashworths And it is ordered that he together with the Male Labouring Tithables Convenient thereto do forthwith Clear and keep the same in Repair According to Law.

2 October 1751 O. S., Page 474
John Ashworth is appointed Surveyor of the Road leading from John Ashworths to the fork of Allens Road. And it is ordered that he together with the Male labouring Tithable Persons Convenient thereto do forthwith Assist the said John in Clearing and keeping the same in Repair According to Law.

2 October 1751 O. S., Page 474
John Winningham is appointed Surveyor of the Road Leading from the fork of Allens Creek to Winninghams Road. And it is ordered that he together with the Male Labouring Tithable Persons Convenient thereto do forthwith assist the said John in keeping the same in Repair according to Law.

2 October 1751 O. S., Page 475
Richard Womack is ordered together with the Assistance Convenient thereto, to lay open & Clear a Road the Best and Most Convenient way from the old Road that Crosses Little Roanoake near Thomas Nashes, and that he keep the Same in Repair According to Law.

2 October 1751 O. S., Page 475
Order'd that Godfry Jones, James Easter and George Foster or any two of them being first sworn &c do diligently view and Examine the way from the old Road near Thomas Nashes to the Church, and Report to the next Court Conveniency &c.

3 October 1751 O. S., Page 488
William Lawson is ordered (together with the following Assistance Convenient thereto) do lay open & Clear a Road the Best and Convenient way from Lawsons Mill to Boyds Ferry and that he keep the same in Repair According to Law

3 October 1751 O. S., Page 488
Robert Wynn is ordered (together with the following Assistance Convenient thereto) do lay open and Clear a Road the best and Convenient way from Banister Bridge to Wades Ferry, And that he keep the same in Repair According to Law

3 October 1751 O. S., Page 488
David Gwin is appointed Surveyor of the Road leading from Samuel Perrins to Captain Bouldins And it is ordered that he together with the following assistance, to wit, the Male Labouring Tithable persons of the Said David Gwin & Captain Bedfords hands do forthwith Clear and keep the same in Repair according to Law

3 October 1751 O. S., Page 488
Order'd that Cornelius Cargill William Caldwell & Thomas Bouldin and James Mitchell or any Three of being first Sworn & c do View and Examine the way called Wittons Road and Twittie's a Cross Meherrin and Report to the next Court the Conveniency & or Inconveniency thereof, And it is further ordered that all the Persons that formerly worked on the said Roads be Discharged from working on the same.

3 October 1751 O. S., Page 500
Order'd that a bridge be Errected and built over the Middle Meherrin River where Cox's Road crosses the Said River, And Abraham Martin & Thomas Bouldin Gent. are appointed & Desire'd to lett to undertakers for Seven Years & to be Built in Such manner and for as their Descression Shall think fitt & Convenient --

3 October 1751 O. S., Page 503
James Taylor is appointed Surveyor of the Road leading from Reedy Creek to the fork of Slate Rock Road And it is ordered that he together with all the assistance Convenient thereto do forthwith Clear & keep the same in Repair According to Law.

3 October 1751 O. S., Page 504
William Blagrave is appointed Surveyor of the Road leading from Middle Meherrin to the North Meherrin. It is ordered that he together with the Hands Convenient thereto doth forthwith Clear & keep the same in Repair According to Law.

3 October 1751 O. S., Page 504
On the Motion of James Stewart Licence is granted him to keep a Ferry on Stanton River at his own Landing he giving Security where upon he together with [blank in book] his Securities Entered into & acknowledged their Bonds According to Law for that Purpose.

And the Court do adjudge that he take Three pence Three farthings for a Man & Seven pence half penny for man & horse and woman

3 October 1751 O. S., Page 505
Mathew Marable is appointed Surveyor of the Road leading from his Store to the Court house, and It is ordered that he togeth with the Hands Convenient thereto do forth with Clear & keep the same in Repair According to Law.

4 October 1751 O. S., Page 506
William Wilson is appointed Surveyor of the Road leading from Meherrin to Bears Element, And it is ordered that he together with all the Assistance Convenient thereto do forthwith Clear & keep the same in Repair according to Law

4 October 1751 O. S., Page 507
John Williams is appointed Surveyor of the Road leading from Bears Element to the fork of the Road at flat Rock And it is ordered that he he together with all the Assistance Convenient thereto do forth with Clear & keep the the same in Repair According to Law

4 October 1751 O. S., Page 518
Stephen Mallet is appointed Surveyor of the Road whereof Francis Elledge was late Surveyor And it is ordered that he together with the Assistance that assisted the said Elledge on the said Road do forth with Clear & keep the same in Repair According to Law

4 October 1751 O. S., Page 518
Joseph Greer is appointed Surveyor of the Road leading from Deloneys old ordinary over all the Branches of Meherrin into the Old road, and it is ordered that he together with all the Male Labouring Tithable Persons Convenient thereto do forthwith Clear & keep the same in Repair according to Law.

15 November 1751 O. S., Page 520
County Levy
To William Caldwell for Building a Bridge over Turnip Creek ...
    11.14.0

***

To Silvanus Walker for Building a Bridge over the North Meherrin River ...
    30..0..0
To Michael Johnson for Building a Bridge over Ledbetter ...
    6..7..6

7 January 1752 New Style, Page 528
William Carby is appointed Survey of the Road leading from Banister Bridge to Wades Ferry and it is ordered that he together with the Male Labouring Tithable Person Convenient thereto do forth with Clear & keep the same in Repair according to Law.

7 January 1752 New Style, Page 529
Joseph Greer is appointed Surveyor of the Road leading from Delonys Ordinary to Mitchells Foard & to the Fork of Wiltons road and it is ordered that he together with all the Convenient Persons thereto do forth with Clear & keep the same in Repair according to Law

7 January 1752 New Style, Page 529
Jeremiah Hatcher is appointed Survey of the road from Wiltons fork over the North Meherrin to the old Road and it is ordered that he together with the following Assistance, to wit, Philip Pondexter & his hands, John Twitty & his, John Chandler Phillip Cockrum, Jeremiah Hatcher & his William Mackadieu, Daniel Haynes, or whoever lives on his Place, Nicholas Hobsons and his, John Hawkins, Richard Hawkins, John Hobsons & his hands at Masons Quarter, do forth with Clear and keep the same in Repair According to Law

7 April 1752 New Style, Page 8
Ordered that the Collector of this County do pay Abraham Cocke Gent his Proportionable Part of the Expences in Building a Bridge over Nottoway River Part in this County.

7 April 1752 New Style, Page 11
Ordered that the Collector of this County do Pay Abraham Cock Gent the sum of Six Pounds Six Shillings & Eleven Pence one Farthing for this Countys Proportion for the Bridge built over Nottoway River.

7 April 1752 New Style, Page 13
Order'd that McNess Good, Charles Talbot & Abra Martin View and Mark the most Convenient Way from the Mossing ford on Little Ronoke into M^r. Martins Road and Report to the Next Court the Conveniency thereof.

7 April 1752 New Style, Page 15
Joseph Johnson is appointed Surveyor of the Road Leading from the moth of the Finnywood Creek to the fork of the Roberson Creek, And it is Ordered that all the male Labouring tithable Persons Convenient thereto forthwith Assist in Clearing and Keeping the same in Repair According to Law.

7 April 1752 New Style, Page 15
James Breedlove is appointed Survey^r. of the Road leading from the fork of Roberson Creek to Breedloves Creek, And it is ordered that all the Male Labouring tithable Persons Conven thereto Assist in Clearing and Keeping the same in Repair According to Law.

7 April 1752 New Style, Page 15
George Wells is appointed Surveyor of the Road Leading from Breedloves Creek to Owls Creek. And it is Ordered that he together with all the Male Labouring tithable Persons Convenient thereto forthw^h. Clear and keep the same in Repair According to Law.

7 April 1752 New Style, Page 16
Francis Moore Petty is Appointed Surveyor of the Road leading from Owls Creek to Moores Road, And it is Ordered that he together with all the Male labouring tithable Persons Conven thereto forthwith Clear and keep the same in Repair According to Law.

7 April 1752 New Style, Page 16

The Petition of Sundry Inhabitants of this County Setting forth the Inconveniency they Labour under for want of a Road, And the Convenience of a Road that might be laid off out of Roanoke Road at the head of the Ready branch into Williams's Road near the head of Crooked Creek, is granted And According to the Prayer of the s^d. Pet^o.

it is Ordered that Jeffery Russell or Charles Weatherford be Surveyor thereof and that the following hands, to wit, John Tomson Abraham Talley, W^m. Weatherford, Henry Talley, William Traylor, John Talley, Peter Parish, X^r. Johnson Jno. Weatherford Samuel Young Richard Tompson Jeffery Russel & Charles Weatherford forthwith lay open and keep the same in Repair According to Law.

7 April 1752 New Style, Page 17
On the Petition of Sundry Inhabitants of this County, It is ordered that a Bridge be built over Louse Creek where the Road crosses the same, and that W^m. Caldwell Gent & James Hunt agree with workmen to build the same in such maner and upon such Terms as they Shall think Proper.

7 April 1752 New Style, Page 18
William Stone is appointed Surveyor of the Road whereof Rice Price was formerly survey^r of, And its ordered that all the male Labouring Tith^a. Persons Convenient thereto forthwith assist in Clearing and keeping the same in Repair According to Law.

8 April 1752 New Style, Page 28
On the Petition of James Mitchell Gent Licence is Granted him to keep a Ferry in this County from his Land to the Land of one Thomas Anderson, he giving Security, Whereupon he together with Leonard Claiborne Gent his Security entered into and Acknowledged their Bond for that Purpose, And the Court do adjudge that he shall take three pence three farthings ferryage for a Man & the same for a Horse & no more.

5 May 1752 New Style, Page 33
Thomas Winningham is appointed Surveyor of the Road Leading from the fork Randolphs Road to the Bridge over Maherrin River near Winninghams And it is ordered that he together with all the male Labouring Tithable Persons Convenient thereto forthwith Clear and keep the same in Repair According to Law.

5 May 1752 New Style, Page 33
Daniel Malone is appointed Surveyor of the Road from the Bridge over Maherrin River Near Winninghams to the Fork of the Road below Stokes's, And it is ordered that he together with all the Male Labouring tithable Persons Convenient thereto forthwith Clear and keep the same in Repair According to Law.

5 May 1752 New Style, Page 37
Order'd that a Road be laid off and Cleared the best and most Convenient way from Well's Race Paths to go by Delonys Ordinary into the Road at Mizes Ford, And that Dennis Lark be Surveyor of the upper part thereof, and Samuel Homes Surveyor of the Lower part, And that all the male labouring Tithable Persons Convenient thereto forthwith Assist in Clearing and keeping the same in Repair According to Law.

5 May 1752 New Style, Page 38
Samuel Perrin is appointed Surveyor of Randolphs Road from the upper fork thereof to Capt Bouldins And it is ordered that he together with the Following Male Labouring Tithable Persons, to wit, Bedfords Overseer and the male Labouring Tithables under his Care, Joel Towns, the widow Gwins hands David Gwin & his, John Francis, James Brumfeld, Joseph Perrin & his & Thomas Bedfords, forthwith Clear and keep the same in Repair According to Law.

5 May 1752 New Style, Page 38
Ordered that the Collector of this County do Pay John Speed Twelve Pounds Current Money for a Bridge by him built over Allens Creek in this County.

5 May 1752 New Style, Page 38
Ordered that the Collector of this County do Pay John Cox Nine Pounds five Shillings Current Money for a Bridge by him built over the Roberson Creek in this County.

5 May 1752 New Style, Page 39
Edward Powel is appointed Surveyor of the Road leading from Banister Bridge to Wades Ferry And it is Ordered that he Together with all the Male Labouring Tithable Persons Convenient thereto forthwith Clear and keep the same in Repair According to law.

2 June 1752 N. S., Page 43
Ordered that Stephen Bedfords Overseer at his Quarter in this County with Male Labouring Tithables under his Care, Godfrey Jones & his Male Labouring tithables, James Easter & his, John Glascock & his, Charles Sullivant & his, Nathaniel Terry & his, Josias Randle & his, James Sullivant & his Man Leonard Ashworth and all others Convenient do Assist Thomas Bouldin & Clement Read Gent Surveyors of the Road from the Bridge to Randles Road in Clearing the same and making a Causeway.

2 June 1752 N, S., Page 44
Ordered that a Road be laid off and Cleared the best and Most Convenient Way from Stewards Ferry on Stanton River to the Mossing ford on Little Ronoke River, And William Redman is Appointed Surveyor thereof, And it is ordered that he together with all the Male Labouring tithable Persons Convenient thereto forthwith Clear and keep the same in Repair According to Law.

2 June 1752 N. S., Page 46
Ordered that Thomas Staples & his male Labouring Tithables, John Ingrum & his, John Winningham, Jonathan Davis & his, John Nance & his, James Roberts & his, Thomas Rutherford, Thomas Nance, Richard Nance, William Nance, Cornellius Cranshaw, Patrick Tilen, Barnaby Wells, Argil Blaxton do Assist Thomas Winningham Surveyor of the Road leading from the fork of Randolps Road to the Bridges over Maherin River near Winninghams, in Clearing and keeping the same in Repair According to Law.

2 June 1752 N. S., Page 47
On the Motion of Charles Talbot and for Reasons appearing to the Court, it is Ordered that Joseph Perrin View the Road from the Mossing ford into Martins Road in the Room of the said Charles Talbot & Report to the next Court the Convenience thereof.

2 June 1752 N. S., Page 48
William Stone is appointed Surveyor of the Road leading from the Fish dam of Otter River to James Johnsons at the Poplar Spring And 'tis Ordered that he together with the Male Labouring Tithable Persons, to wit, Mathew Talbot, John Anthonys hands, John Paynes at his Quarter, Rice Price, Thomas Price, Edward Mobberly, Benjamin Mobberly, John Mobberly, Thomas Pitman, William Moss, William Bramlet, Thomas Branch & John Turner and their Male Labouring tithables, forthwith Clear and keep the same in Repair According to Law.

2 June 1752 N. S., Page 52
James Mitchell, William Caldwell & Thomas Bouldin Gent having on Oath Reported to the Court that it will be more Convenient & Usefull to the Public to Lay open a Road from out of the Road called Twittys Road from where Wittons Road intersects the same the most convenient way into Eledges Road on the south Side of Maherrin River, than to Continue the Said Twittys Road from the said Wittons Road across the three Branches of Maherrin River, It is therefore Ordered that a Road be laid off from the said Twittys Road where the said Wittons Road Intersects the same the best and most Convenient way into Elledges Road as aforesaid And that Thomas Charlton be Surveyor thereof and that all the Male Labouring tithable Persons which have usually worked on the said Twittys Road from Wittons Road across the three Branches of Maherrin be Discharged from all further Services thereon, And that they together with all other male Labouring Tithable Persons Convenient thereto forthwith Clear and keep the same in Repair According to Law.

2 June 1752 N. S., Page 55
On the motion of Thomas Hawkins & others ordered that a Bridge built over the North Maherrin River where Elledges Road Crosses the same, And that Richard Witton & Hugh Lawson Gent agree with Workmen to build the same on as cheap Term as they can, with Condition that they Maitain & keep the same in Passable Repair for the space of Seven Years at their own Risque & Charge.

2 June 1752 N. S., Page 59
David Caldwell came into Court and undertook to Support & Maintain & keep in Passable Repair, the Bridge by him Built over Cubb Creek for the Space of Five Years from this Day.

2 June 1752 N. S., Page 60
Ordered that Clement Read Treat with the County Court of Amelia, to Joyn with this County in building a bridge over Nottoway River at John Nights, And Report to the Next Court their Approbation thereof.

2 June 1752 N. S., Page 60
Ordered that Abraham Martin & Corn[s] Cargill Gent View the Bridge built by Memucan Hunt over Blew Stone Creek and Report to the Next Court the Sufficienty thereof.

2 June 1752 N. S., Page 60
On the Petition of William Fuquay, Licence is Granted him to keep a Ferry on Stanton River from his Land to the Land of Mr William Cole, Provided he enter Bond with Sufficient Security in the Clerk office some time between this and the Next Court. And the Court do adjudge that he shall take four Pence Ferriage for a Man & Six Pence for a Man & Horse, and no more.

2 June 1752 N. S., Page 60
Ordered that a Road be laid off and Cleard the best and most Convenient Way from William Saffolds by Freemans Mill on Flat Rock Creek to Kettlestick Creek, And William Saffold is appointed Surveyor thereof from his house to Freemans Mill, and Nicholos Callaham from thence to Kettlestick, And it is Ordered that they together with all the Male Labouring Tithable Persons Convenient thereto forthwith Clear and keep the same in Repair According to Law.

8 July 1752 N. S., Page 148
Ordered that a Road be laid off & Clear'd the best and most Convenient way from that Road of Martin Fifers to Kings ford on Ronoke River, And James Coleman is appointed Surveyor thereof, And it is Ordered that he together with the following male Labouring Tithable Persons, to wit, Thomas Lanear, William Hill, Jeremiah Glaunch, Calub Blackwell, Andrew Tecker, Martin Fifer, Michael Parengame, Michael Hoprick, Henry Sage, Nicholos Major, John Akin & William Bevell with their male Labouring tithable Persons, forthwith Clear & keep the same in Repair According to Law.

8 July 1752 N. S., Page 148
Ordered that a Bridge be Built over the South fork of Maherin River in this County Where the Road Calld Wittons Road Crosses the same, And that Hugh Lawson & Richard Witton Gent Lett the same to undertakers to be built in such manner, with such condition & upon such Terms, as they shall think fit & Convenient.

8 July 1752 N. S., Page 148
Ordered that a Bridge be built over the South fork of Maherrin River where the Road crosses the same Near John Cox's, And that Richard Witton, Mathew Marrable & Abrham Martin G^t. Lett the same to undertakers, to be built in such manner & upon such Terms as they shall think Proper.

4 August 1752 N. S, Page 162
Ordered that William Williams be Summon'd to appear at the Next Court to answer such things as shall then & there by Objected against him of and Concerning a Brigde by him built over Maherin River in this County, and that he do not Depart thence without Leave of this Court.

4 August 1752 N. S, Page 164
Ordered that the Collector do pay George Williams William Gee & William Edwards the sum of Ten Pounds Current Money for Building a Bridge over the North Maherrin River in this County.

4 August 1752 N. S, Page 165
Ordered that Hutchins Burton George Holloway, Henry Howard & John Speed (being first Sworn before a Majistrate of this County for that Purpose) do View the Way to the Ferry Petition'd for by William Abbot, & the way to Jeffersons Ferry and Report which is the most Convenient way, here to the Next Court.

4 August 1752 N. S., Page 173
On the Motion of Hampton Wade, Henry Roberson is appointed Surveyor of the New Road appointed by an Order of this Court, from John Cargills, the best and most Convenient way into Cox's Road, And that he together with the Following male Labouring Tithable Persons, to wit, Cornelius Cargill & his male Labouring Tithable Persons, John Cargill & his, Samuel Harris & his, Colo. Byrds Overseer & the hands under his Care William Perry & his, William Roberts & his, Those of the Estate of Tandy Walker Decd., George Mosly & his & Robert Wade S^r. & his, forthwith Clear and keep the same in Repair According to Law.

4 August 1752 N. S., Page 173
Ordered that Scherer Degrafenreidt & George Walton do meet the Persons appointed by the County Court of Amelia for that Purpose, And Lett a Bridge formerly ordered to be built over Nottoway River to Undertakers upon such Terms as they Shall Proper.

5 August 1752 N. S., Page 207
Ordered that the Collectors do Pay Memucan Hunt the sum of Fourteen Pounds two Shillings Current Money, for a Bridge by him Built over Blew Stone Creek in this County.

5 August 1752 N. S., Page 221
Samuel Moreton is appointed Surveyor so much of the Road that leads from the upper Bridge on Little Roanoke River, to Charles Andersons, or shall be in this County, And it is Ordered that he together with the Following male Labouring tithable Persons, to wit, the male Labouring Tithables of him the said Samuel Morton, Robert Childress, Charles Harris, Owen Sullivant, William Cook, Christopher Gormer, & Philip Hudgins forthwith Clear & keep the same in in Repair According to Law.

5 August 1752 N. S., Page 221
Reuben Morgin is appointed Surveyor of the Road Whereof Julius Nichols was Surveyor, And it is Order'd that he together with all the male Labouring Tithable Persons Convenient thereto forthwith clear & keep the same in Repair According to Law.

1 September 1752 N. S., Page 231
On the Motion of William Marrable & others, It is Ordered that a Road be laid off & Clear'd from Mathew Marrables Crossing the South fork of Maherrin River above & Near Millers Quarter & from thence into John Cox's Road, And it is further Ordered that William Harris & McNess Goode do Search & Examine for the best & most Convenient Way for such Road and make Report to the Next court.

1 September 1752 N. S., Page 247
William Williams appear'd on being Summon'd, by an order of the last Court, to answer such things a should be Objected against him of and Concerning a Bridge by him built over Maherrin River in this County, and thereupon it is ordered that the Clerk do Bring to the Next court a Bond Enter'd into by the said William, (And the former Order of Court Concerning the said Brigde, And that the said Summons be Continued, until the Next Court.

3 September 1752 N. S., Page 273
At a Court Continued and held for Lunenburgh County on their fourteenth day of September, being the third day of Septr. according to the Old Stile, in the XXVIth Year of the Reign of our Sovereign Lord King George the Second and in the Year of our Lord God one thousand seven hundred & fifty two.

Note: See p. v. This is the correction for the day of the month according to the New Style Calendar begun on January 1, 1752.

14 September 1752 N. S., Page 287
Ordered that the Male Labouring Tithable Persons at William Bookers Quarter, those of Anothony Walke, William Gee, & William Wilson, be added to the hands that work on the Road Call'd Wittons Road, And that the Surveyor thereof, together with the hands aforesaid, do forthwith Clear and keep the same in Repair According to Law.

14 September 1752 N. S., Page 287
Ordered that a Bridge be built over Little Ronoke where the Road Crosses the same near Clement Read's, And that Clement Read & Thomas Bouldin Lett the same to Undertakers, upon the Cheapest Terms that they Can, to be Warrented Seven Years, And make Report to the next Court of their Proceedings thereon.

3 October 1752 N. S., Page 289
Order'd that the Road call'd Twittys old Road be Continued a Road, that .... John Twitty be Surveyor thereof, And that he together with the Male Labouring Tithable Persons following, to wit, those of Philip Poindexter, those of the said John Twitty, Joseph Davis, William Lax, Philip Cockerham, & John Chandler forthwith Clear & keep the same in Repair According to Law.

3 October 1752 N. S., Page 289
Mathew Marrable Gent is appointed Surveyor of the Road leading from Kings Road by his house to this Courthouse, And it is Ordered that he together with all the Male Labouring Tithable Persons Convenient thereto, forthwith Clear & keep the same in Repair according to Law.

3 October 1752 N. S., Page 290
Daniel Malone is appointed Surveyor of that Part of Willinghams Road; from Maherrin River to Where James Roberts's Waggon Road comes into the same, And it is Ordered that he together with all the Male Labouring Tithable Persons Convenient thereto, forthwith Clear & keep the same in Repair According to Law.

3 October 1752 N. S., Page 290
Richard Stokes is appointed Surveyor of the Road called Winninghams Road, from where James Roberts's Waggon Road comes into the same to the bridge Near Hampton Wades, And it is Ordered that he together with all the Male Labouring Tithable Persons Convenient thereto forthwith Clear & keep the same in Repair According to Law.

3 October 1752 N. S., Page 300
George Farron is appointed Surveyor of the Road leading from Jeffersons Ferry, to the County Line, And it is ordered that he together with all the Male Labouring tithable Persons Convenient thereto, forthwith Clear & Keep the same in Repair According to Law.

3 October 1752 N. S., Page 303
Field Jefferson Gent having Obtain'd, an order of this Court to Keep a Ferry Over Roanoke River; Came into Court & he together with David Stokes Gent his Security Enterd into & Acknowledged their Bond for that Purpose.

3 October 1752 N. S., Page 303
Field Jefferson, Cornelius Cargill, & Richard Witton Gent or any two of them are appointed & Desired to treat with Persons to undertake the building a Bridge over Maherrin River at Mizes ford on the terms & Conditions following, that is to say, the Undertakers are to Suppor & Maintain the bridge in Passable Repair for the Term of Seven Years rebuild the same if carry'd away in the time or refund the Money they shall have for building the same, as the Contractor Shall thing fitt, for the Performance of which the Undertakers are to give the Contractors bond with Sufficient Security.

3 October 1752 N. S., Page 304
Andrew Mc.Connel is appointed Surveyor of the Road from Reedy Creek to the Musterfield Branch, And it is Ordered that all the Male Labouring Tithable Persons Convenient thereto forthwith assist in Clearing & keeping the same in Repair According to Law.

3 October 1752 N. S., Page 304
Daniel Hayse is appointed Surveyor of the Road Call'd Reedy Creek Road from the Musterfield Branch to the fork of the same, And it is Ordered that he together with the Male Labouring tithable Persons Convenient thereto forthwith Clear & keep the same in Repair According to Law.

3 October 1752 N. S., Page 304
For Reasons Appearing to the Court, it is Ordered that William Abbots Ferry over Ronoke River be Suppress'd.

3 October 1752 N. S., Page 305
On the motion of William Marrable leave is Granted him to lay open a Road from his house into the Courthouse Road, With his own hands, Provided that Whenever there shall be any Obstruction to the same that he will immediately turn the same.

7 November 1752 N. S., Page 309
Ordered that the Road Calld Reuben Morgans Road be Continued a Road, and that the Said Reuben Morgan be Surveyor thereof, and that he to together with all the Male Labouring Tithable Persons Convenient thereto forthwith Clear & keep the same in Repair According to Law.

7 November 1752 N. S., Page 309
Samuel Manning is Appointed Surveyor of the Road leading from flat Creek to Thomas Wells's, and it is Ordered that he together with the Male Labouring Tithable Persons Convenient thereto forthwith Clear & keep the same in Repair According to Law.

7 November 1752 N. S., Page 316
Ordered that McKness Goode, do Continue Surveyor of the Road Whereof he is Now Surveyor

7 November 1752 N. S., Page 316
Ordered that Mathew Marrable Gent do Continue Surveyor of the Road whereof he is Now Surveyor.

7 November 1752 N. S., Page 316
Ordered that Abraham Martin Gent do Proportion and set apart the Number of Male Labouring Tithable Persons Convenient to the Road Whereof Mackness Goode & Mathew Marrable are Surveyors in such manner as he shall think fit, between the said Surveyors.

7 November 1752 N. S., Page 317
Grand Jury Presentments
... James Cocke for not keeping the Road in good Repair Whereof he is Surveyor...
... Steph. Mallet & William Williams for not keeping or Repairing the Bridge over Cox Creek at Col^o. Ruffins Mill ...

* * *

... the Overseer for not Keeping the road in good Repair from the Court house to Mathew Marrables ...

* * *

... the Oversser of the Road from the Court house to Delonys old Ordinary ...

7 November 1752 N. S., Page 321
John Philip Weaver is appointed Surveyor of the Road leading from Goose Creek to the Extent of the County upwards, And it is Ordered that he together with all the Male Labouring Tithable Persons Convenient thereto forthwith Clear and keep the same in Repair According to law.

7 November 1752 N. S., Page 321
Peter Holland is appointed Surveyor of the Road Leading from Goose Creek to Little Otter River, And it is Ordered that he together with all the Male Labouring Tithable Persons Conven^t thereto forthwith Clear and keep the same in Repair According to Law.

7 November 1752 N. S., Page 322
John Eckols is appointed Surveyor of the Road Leading from Little Otter River to great Otter River, And it is Ordered that he together with all the Male Labouring Tithable Persons Convenient thereto forthwith Clear & keep the same in Repair According to Law.

7 November 1752 N. S., Page 322
Edward Mobberly is appointed Surveyor of the Road Leading from Otter River to the Fish Dam Road near the Church, And It is ordered that he together with all the Male Labouring Tithable Persons Convenient thereto forthwith Clear & keep the same in Repair According to Law.

7 November 1752 N. S., Page 323
Daniel Hayse is Appointed Surveyor of the Road whereof Robert Leveret was formerly Surveyor, And it is Ordered that he together with all the Male Labouring Tithable Persons Convenient Thereto forthwith Clear & keep the same in Repair According to Law.

8 November 1752 N. S., Page 332
Grand Jury Presentments
... the Overseer of the Road from Reedy Creek Church to Richard Stoke's Road for not Setting up Posts of Directions According to Law ...
... the Overseer of the Road from the Court house to Capt. Cox's at the forks of the same ...

8 November 1752 N. S., Page 360
Grand Jury Presentments
... the Overseer of the Road from Reedy Creek Church to Richard Stokes's Road for not setting up Posts of Directions According to Law ...
... the Overseer of the Road from the Court house to Capt Cox's at the forks of the same ...

9 November 1752 N. S., Page 391
David Stokes and Lyddal Bacon Gent are appointed and Desired to, appoint Surveyors & hands to work on the Road leading from the Road Call'd Hampton Wades Road to Randolphs Road, and Likewise of the Road leading from Waltons Road to this Courthouse, as they shall think fit & Convenient.

9 November 1752 N. S., Page 391
Ordered that John Cox lengthen & Repair the Bridge by him built over Maherrin River, and when the same is Completed, The said John Cox and this Court are to chuse two Persons to Value the same, and they Cant Agree they two chuse a third Person and make Report to this Court.

5 December 1752 N. S., Page 395
Abram Martin Gent & William Jones are appointed Surveyors of the Road from Randolphs Road to Kings Road, And the said Abram Martin is appointed & Desired to Devide the said Road & the Male Labouring Tithable Persons Convenient thereto, between himself & the said Jones, as he shall think Proper.

5 December 1752 N. S., Page 395
William White is appointed Surveyor of the Road Whereof Joseph Greer was formerly Surveyor, And it is Ordered that he together with all the Male Labouring Tithable Persons Convenient thereto, forthwith Clear and keep the same in Repair According to Law.

5 December 1752 N. S., Page 395
Joseph Granger is appointed Surveyor of the Road Whereof Abraham Cook was formerly Surveyor, And it is Ordered that he together with all the Male Labouring Tithable Persons Convenient thereto forthwith Clear & keep the same in Repair according to Law.

5 December 1752 N. S., Page 395
John Colbreth is appointed Surveyor of the Road Leading from Aarons Creek to Palmers ford, And it is Ordered that he together with all the Male Labouring Tithable Persons Convenient thereto, forthwith Clear & keep the same in Repair According to Law.

5 December 1752 N. S., Page 404
Samuel Ashworth is appointed Surveyor of the Road Leading from the said Ashworths to Peter Hudsons foard, on Stanton River, and it is Ordered that he together with all the Male Labouring Tithable Persons Convenient thereto forthwith Clear and keep the same in Repair According to Law.

5 November 1752 N. S., Page 404
James Cocke is appointed Surveyor of the Road leading from Hudsons foard to Randolphs Road, and it is Order'd that he together with all the Male Labouring Tithable Persons Convenient thereto forthwith Clear & keep the same in Repair According to Law.

6 December 1752 N. S., Page 409
Ordered that Field Jefferson Gent do appoint hands to Assist John Humphris in Clearing the Road Whereof he is Surveyor, as he shall think Proper.

6 December 1752 N. S., Page 409
County Levy
To John Austin for Removing and Repairing Little Ronoke Bridge, and taking care of and Securing Wards fork Bridge in the time of Several Freshes ...   300

* * *

To William Caldwell Gent for building a Bridge over Louse Creek ...   9.16.0

* * *

To Dennis Lark for two Sign Boards   50 (lbs. of tobo.)
To Thomas Wynne for two Sign Boards   50

* * *

To Abraham Martins Account Allow'd for a Bridge over Little Ronoke   22.0.0

* * *

To Robert Wade Sen for the Use of his Ferry Boat   500
To Richard Witton and Hugh Lawson Gent for two Bridges over Maherrin River .. 59.0.0
To Richard Witton and Mathew Marrable Gent for a Brid Near John Cox's in this County ...   29.19.0
To Richard Witton, Cornelius Cargill and Field JeffersonGent for a Bridge at Mizes foard ...   97.0.0

6 February 1753 N. S., Page 447
Robert Woods is appointed Surveyor of the Road leading from Coles Road to Little Ronoke Bridge Near Clement Reads, And it is Ordered that he, together the following male Labouring Tithable Persons, to wit, those of the Widow Gwin, Osborne Keeling, Abraham Lunderman, John Lee, John Mount, Richard Treadway, Samuel Johnson, William Johnson, Daniel Humphris, John Austin & his sons, James Mc.Glaughlin, Wilson Matox, John Atkins & his sons, Marmaduke Stanfield, & William Stanfield &

their Hands, William Covington & his, Stephen Collins & his, And John Cooke do forth with Clear and keep the same in Repair According to Law.

6 February 1753, N. S. Page 450
Joseph Cranshaw is appointed Surveyor of the Road leading from Byrds Road to the Mouth of Butchers Creek, And it is Ordered that he, together with the following Male Labouring Tithable Persons, to wit, John Fletcher Francis Bray & his male Labouring Tithables Mathew Tanner & his Gideon Chranshaw & his Valentine Mullins & his, John Wilkins, James Wilkins, Thomas Wilkins, Thomas Hawkins & his, those of Mr. George Currie, Thomas Satterwhite & his, & Thomas Tate, do forthwith Clear and keep the same in Repair According to Law.

6 February 1753 N. S., Page 451
Thomas Anderson is appointed Surveyor of the Road Leading from Byrds Road to this Courthouse, And it is Ordered that he together with all the Male Labouring Tithable Persons Convenient thereto forthwith Clear and keep the same in Repair according to Law.

6 February 1753 N. S., Page 465
Thomas Jones is appointed Surveyor of the Road leading from Randolps Road to the Horse Pen Creek, And it is Ordered that he together with all the Male Labouring Tithable Persons that formerly worked under Abraham Martin on the said Road, do forthwith Clear and keep the same in Repair According to Law.

7 February 1753 N. S., Page 511
Thomas Satterwhite is appointed Surveyor of the Road leading from Palmers landing on Butchers Creek, To Martin ffifers Road And it is Ordered that he together with the following male Labouring Tithable Persons, to wit, those of the said Thomas Satterwhite, Richard Palmer & his, Thomas Hawkins & his, Mathew Tanner & his, John Clarke & his, forthwith Clear & keep the same in Repair According to Law.

7 February 1753, N. S., Page 511
Order'd that a Road be laid off and Clear'd the best and most Convenient way from the mossing ford on Little Ronoke River, to James Stewarts Ferry, And that Charles Talbot be Surveyor thereof, And that he together with all the male Labouring Tithable Persons Convenient thereto forthwith Clear and keep the same in Repair according to Law.

6 March 1753 N. S., Page 532
On the Petition of John Speed and others, It is Ordered that a Bridge be built over Miles Creek near the Church call'd Ronoke Church in this County, And it is Ordered that George Baskervile and John Speed treat with Workmen to build the same on such terms in such manner and for what Consideration they shall think proper and that the undertaker do warrent & Maintain the same Seven Years.

6 March 1753 N. S., Page 533
Thomas Jones is appointed Surveyor of the Road leading from Twittys Creek to M^r. Thomas Bouldins, and also of the Court house Road from the Horse Pen Creek to Bouldins Road, And it is Ordered that he Together with all the Male Labouring Tithable Persons convenient thereto forthwith Clear and keep the same in Repair according to Law.

6 March 1753 N. S., Page 534
Richard Witton Gent is appointed Surveyor of the Road whereof Jeremiah Hatcher was formerly Surveyor, And it is Order'd that William Gee & his Male Labouring Tithable Persons & those at William Bookers Quarter do forthwith assist in Clearing and keeping the same in Repair According to Law.

6 March 1753 N. S., Page 534
Ordered that William Harris and John Flin being first sworn before a Majistrate of this County according to Law, do View Examine and Search for the best and most Convenient way for a Road to be laid off and Clear'd from the Middle fork of Blew Stone Creek to this Courthouse and that they Return a Report thereof here to the Next Court.

6 March 1753 N. S., Page 539
On the motion of David Caldwel [part of page missing] and others, it is Ordered that a [part of page missing] Bridge be built over Wards fork Creek in this County, that Clement Read and William Caldwel Gent treat with workmen to Build the same upon such place in what manner for what Consideration they shall think proper and that the undertaker do warrant and Maintain the same Seven Years.

3 April 1753 N. S., Page 581
Joel Chandler is appointed Surveyor of the Road in the room of Joseph Chandler And it is ordered that he together with the hands that Worked on the said Road under the said Joseph Chandler, do forthwith Clear & keep the same in Repair According to Law.

3 April 1753 N. S., Page 581
On the motion of William White, It is Ordered that the following hands, to wit, Christopher Hudson & his two Negro's, John Humphris, James Tucker, William Bevil, Adam Winders & his Negro, James Bilbo, John Carrel, William Jones & his two Negro's Joseph Ragsdales Negro, Warner Tucker, Joseph Greer & his Prentice, Edward Good & his Negro, Joseph Gill & his, & Robert Hatcher, do assist the said White in Clearing & keeping in Repair the Road whereof the said White is Surveyor.

3 April 1753 N. S., Page 582
On the motion of John Cooper & others it is Ordered that the Road Cleared by William Booker along the Ridge between Crooked Creek & Bairs Element into the Road that leads by Jennings's Ordinary, and from the Little Beaver Pond into the said Bookers Road be Continued a Road And William Saffold is appointed Surveyor thereof, And it is Ordered that he together with the following male Labouring tithable Persons, to wit, those of the said William Booker at his Quarter on Masons Creek & on the Little Beaver Pond, those of M^r. John Booker, John Cooper & his do forthwith Clear & keep the same in Repair According to Law, And it is ordered that the said Hands be exempt from other Roads.

3 April 1753 N. S., Page 582
David Garland is appointed Surveyor of the Road Whereof William Saffold was late Surveyor, and it is Ordered that he together with the hands which worked under the said Saffold on the said Road do forthwith Clear & keep the same in Repair according to Law.

4 April 1753 N. S., Page 26
Julius Nichols is appointed Surveyor of the Road Whereof John King was late Surveyor and it is Ordered that he together with the hands that worked under the said King on the said Road do forthwith clear & keep the same in Repair According to Law.

4 April 1753 N. S., Page 31
Richard Ship is appointed to Value the two Additions made by John Cox Gent to a Bridge by him formerly built over Maherrin River, in the Room of Hutchins Burton &c

1 May 1753 N.S., Page 71
Pursuent to a former Order of this Court, Abraham Martin & Richard Ship who were appointed by the said Order to View & Value the additions made by John Cox Gent to the bridge by him formerly built over Maherrin River in this County, this day Return'd their Report thereupon, which is ordered to be Recorded.

1 May 1753 N.S., Page 112
Grand Jury Presentments
...Stephen Mallet for not repairing the bridge over Cocks Creek at delonys Mill ...
* * *
...Andrew Martin Surveyor of a Road ...

1 May 1753 N. S., Page 115
On the motion of Thomas Jones, It is Ordered that the following male Labouring Tithable Persons to wit, James Waldun, those of Clement Read at his Lower Quarter, Joel Towns, John Handcock those of Mary Gwin, those of M^r. Thomas Bedford, Stephen Bedford & his, Joseph Perrin & his, Samuel Perrin, William Perrin, Richard Sansun, Owen Sullivant, & William Fulsher do assist the said Thomas Jones in Clearing & keeping in Repair the Road whereof the said Thomas is Surveyor.

3 May 1753 N. S., Page 127
James Hunt is appointed Surveyor of the Road whereof George Abney was late Surveyor, And it is Ordered that he together with all the Male Labouring Tithable Persons Convenient thereto do forthwith Clear & keep the same in Repair According to Law.

3 May 1753 N. S., Page 127
Order'd that a Road be laid off and Clear'd the best and most Convenient Way from the Mossing foard on Little Ronok River to a Road Called Martins Road, And Charles Talbot is appointed Surveyor thereof, And it is Ordered that he together with all the male Labouring Tithable Persons, Convenient theereto do forthwith Clear & keep the same in Repair According to Law.

3 May 1753 N. S., Page 128
Stephen Mallet Surveyor of a high Way in this County having been Presented for not keeping in Repare a bridge over Cocks Creek in this County, And not appearing to make his defence, It is Consider'd that he make his fine with our Sovereign Lord the King by the Payment of fifteen Shillings Current Money, to his Majestys use And that he Pay the Costs of these Proceedings, And it is Ordered that he be taken &c

3 May 1753 N. S., Page 128
Mathew Marrable Gent Surveyor of the high Way from this Court house to Marrables Ordinary in this County having being Presented for not keeping the same in Repair, Appeared and on hearings as well as the argument of the attorney for our Lord the King, as the said Marrable, It is considered by the Court that the said Presentment be Dismis'd And that the said Marrable go thereof hence without day.

5 June 1753 N. S., Page 130
Nathaniel Christian is appointed Surveyor of the Road leading from the Ridge Joyning the County of Amelia to little falling River, And it is Ordered that he together with the following male labouring Tithable Persons, to wit, Robert Richey, John Richey, Thomas Harvie & his, William Owl, John Marshal, Daniel Burn, William davis & his, Timothy McDaniel, George Lovel, William Almon, John Ussery, John Weatherford, John Wood, James Euin, John Love, Samuel Moon, Thomas Willis, John Mitchell, & John Upcott do forthwith Clear & keep the same in Repair According to Law.

5 June 1753 N. S., Page 166
It appearing to the Court by many years Experience that Maherrin River at a Place called Mizes foard in this County is a Very inconvenient Place to build a Bridge and that no bridge can be made there to stand, And that it is a very Convenient way for a Road to trade, It is therefore Ordered that the Banks of the said foard be Cut down, And that a Flat be built and kept at the Place to Carry over Tobacco &c And Henry Delony, George Baskervile & Hutchins Burton are appointed to let the same to Undertakers, to be Warrented Seven Years.

8 August 1753 N. S., Page 330
George Holloway is appointed surveyor of the Road leading from Allens Creek Bridge to Miles's Creek, and it is ordered that he together with all the male labouring Tithable Persons Convenient thereto do forthwith Clear & keep the same in Repair According to Law.

8 August 1753 N. S., Page 357
Ordered that James Coleman do instanly lay open and Clear the Road Whereof he is surveyor, Wherein John Davis hath fallen Trees.

4 September 1753 N. S., Page 363
A Petition of Richard Foxx for a Ferry over Ronoke River, for Reasons appearing to the Court is Rejected

2 October 1753 N. S., Page 447
George Vaughn is appointed Surveyor of the Road that Crosses Maherrin River at Mizes foard, from the County Line to Where the same intersects Allens Creek Church Road. And it is Ordered that he, together all the Male Labouring Tithable Persons Convenient thereto do forthwith Clear & keep the same in Repair according to Law.

2 October 1753 N. S., Page 448
On the Petition of Joseph Johnson & others Leave is Granted them to lay open & Clear a Road the Nearest and best Way from Silvanus Stokes's S^r. on the Ridge between between the Junaper and the Middle River of Maherrin down Johnsons Roling road to Silvanus Walkers Bridge at Scotts foard on the north fork of Maherrin River, And the said Joseph Johnson is appointed Surveyor thereof

31 October 1753 N. S., Page 467
John Thompson is appointed Surveyor of the road leading from Buck horn to Delonys Old Ordinary, And Henry Delony is appointed & Desired to Nominate Such hands to work on the said Road as he shall think Proper, And it is Ordered that he (together with such hands as the said Delony shall appoint do forthwith Clear and keep the same in Repair According to Law.

31 October 1753 N. S., Page 469
On the Petition of Daniel McNeil & others, Leave is Granted them to Lay open and Clear a Road from the County line near James Yanceys the best and most Convenient Way to William Roysters land on Ronoke & from thence into William Byrds Road. And William Richardson is appointed Surveyor thereof

31 October 1753 N. S., Page 469
John Vance is appointed Surveyor of the Road leading from Otter River to Euings fence, in the room of Adam Baird. And it is Ordered that

he together with all the Male Labouring Tithable Persons Convenient thereto do forthwith Clear & keep the same in Repair According to Law.

31 October 1753 N. S., Page 469
John Mills is appointed Surveyor of the road leading from Euing's fence to the Blew ridge in the room of William Mills, And it is ordered that he together with all the Male Labouring Tithable Persons Convenient thereto do forthwith Clear & keep the same in Repair According to Law.

31 October 1753 N. S., Page 469
James Arnold is appointed Surveyor of the road leading from the Mountain Creek to Buck horn and Henry Delony is appointed & Desired to appoint such hands to assist the said Arnold as he shall think Proper And it is Ordered that the said Arnold (together with such hands as the said Delony shall appoint) do forthwith Clear & keep the same in Repair According to Law.

31 October 1753 N. S., Page 470
On the Petition of William Holt and others, leave is Granted them to lay open and Clear a Road from Little Ronoke Bridge to the head of the south branch of Wards ford Creek, And Alexander Joyce is appointed Surveyor thereof.

31 October 1753 N. S., Page 470
John Anthony is appointed Surveyor of the Road leading from the Fish dam to the Poplar Springs, and it is Ordered that he together with all the Male Labouring Tithable Persons Convenient thereto do forthwith Clear & keep the same in Repair According to Law.

31 October 1753 N. S., Page 470
Stephen Evans is appointed Survey of the Road Call'd Robertsons Road, from Cargills Road to Cox's Road, And John Cox Gent is desired to appoint such hands to work on the same as he shall think Proper, And it is Ordered that the said Evans (together with such hands as the said Cox shall appoint) do forthwith clear & keep the same in Repair According to Law.

31 October 1753 N. S., Page 470
Nathaniel Robertson & Jacob Mitchell are appointed Surveyors of the Road Whereof John Gilliam was late Surveyor. And Henry Delony is appointed and desired to Nominate such hands to work on the said road as he shall think Proper, And it is Ordered that they (together with such hands as the said Delony shall appoint) do forthwith Clear & keep the same in Repair According to Law.

6 November 1753 N. S., Page 479
Thomas Williamson is appointed Surveyor of Robersons Road from Cargills road to Cox's road, And it is Ordered that the hands that work under John Cargill on the lower side of Blew stone Creek, those under John Gorrie & William Hunt, do assist him in Grubing & Clearing the same, and after the same is Open'd, that John Cox Gent do appoint assistance to keep the same in Repair According to Law.

7 November 1753 N. S., Page 481
Joseph Perrin is appointed Surveyor of the Courthouse Road from Randolphs road to the Horse Pen Creek, And Abraham Martin Gent is appointed and Desired, to appoint Persons to work on the same, And it is Ordered that the said Joseph Together with such Assistance as the said Martin shall appoint do forthwith Clear & keep the same in repair According to Law.

7 November 1753 N. S., Page 482
Ordered that John Howel, Mathew Williams, and William Niel, (being first Sworn According to Law) do Search for and find out the best & most Convenient way for a road to be Clear'd from the mouth of the great Branch with runs into Stony Creek to flat rock road, and that they make a report to the Next Court

7 November 1753 N. S., Page 483
On the motion of William Thomas, Licence is Granted him to keep a Ferry over Stanton River about a Mile below the long Island, Opposite to the Ferry Granted him by the County Court of Hallifax, Provided he Enter into Bond with Sufficient Security According to Law, in the Clerks Office some time between this & the Next Court, And the Court do adjudge that he take Six Pence Ferriage for a man for a Man & Horse & no more.

7 November 1753 N. S., Page 483
Ordered that a road be laid off and Clear'd the best and most Convenient Way from Randolphs Road to Thomas's Ferry, And William Steith is appointed Surveyor thereof, And it is ordered that he together with all the male Labouring Tithable Persons Convenient thereto do forthwith lay open Clear & keep the same in repair According to Law.

7 November 1753 N. S., Page 483
John Gorrie is appointed Surveyor of the Road Whereof John Cox Gent was late Surveyor, and it is ordered that he together with such hands as Worked on the said Road under the said Cox do forthwith Clear & keep the same in Repair according to Law.

7 November 1753 N. S., Page 484
William Marrable is appointed Surveyor of the road whereof Mathew Marrable was late Surveyor, And it is Ordered that he together with the hands that Work'd on the said Road under the said Mathew, do forthwith Clear & keep the same in Repair According to Law.

7 November 1753 N. S., Page 484
William Rogers is Continued Surveyor of the road Whereof he hath heretofore been Surveyor, And it is Ordered that William Thompson, William Berry Hill Robert & Daniel Mitchell, John Cunningham & James Cunningham do assist in keeping the same in Repair According to Law.

7 November 1753 N. S., Page 487
Grand Jury Presentments

... the Surveyor of the Road from Robert Wades to this Place ...

* * *

... the Surveyors of the Road from Jennings's to Nottoway for not keeping the same in Repair ...

* * *

... the Surveyor of the Road from Twittys to Jennings's for not setting up Posts of directions at the fork of Wittons Road ...

... the Surveyor of the road for not Setting up a Post of Direction Where Cocks Road leaves Stokes's Road.

7 November 1753 N. S., Page 490
Henry May is appointed Surveyor of the Road Whereof James Cocke was late Surveyor, And it is Order'd that he together with all the male Labouring Tithable Persons Convenient thereto do forthwith Clear & keep the same in Repair According to Law.

6 November 1753 N. S., Page 512
On the Petition of John Wilborne, William Harris, John Roling, Augustine Roling, John Ragsdale, Benjamin Ragsdale & Steven Willis leave is Granted them to lay open and Clear a road the best & most Convenient Way from the Mouth of the Wood Pecker Creek into the road at this Court house, and the said John Wilborne is appointed Surveyor thereof.

5 February 1754 N. S., Page 531

Isaac Hudson is appointed Surveyor of the road Whereof Thomas Williamson was late Surveyor, And it is ordered that he (together with such hands as were by this Court appointed to work on the said road under the said Williamson) do forthwith Clear & keep the same in Repair According to Law.

5 February 1754 N. S., Page 531
Clement Read & Thomas Bouldin are appointed & Desired to View & find out the best & most Convenient Place for a bridge to be Built over little ronoke river at or Near the Mossing Foard, And It is Ordered that they report the Conveniency thereof to this Court

5 February 1754 N. S., Page 540
Daniel Hayse a Surveyor of a High Way in this County, who Stands Presented by the Grand jury for Not keeping the same in repair, appeared & on hearing the arguments on both sides, it is Considered that the Said Presentment be Dismis'd

7 February 1754 N. S., Page 574
Ordered that Thomas Bouldin & James Easter (being first Sworn &c) do View & find out the best & most Convenient Way for a Road to be laid off & Cleard from the old Road Near Thomas Nash's into the road at or near the said Bouldins, And that they make report to the next Court.

5 March 1754 N. S., Page 576
On the Petition of Peter Fontaine Gent Licence is Granted him to keep a Ferry on Stanton River from his own Land, to the Land of James Cocke decd (he giving Security) Whereupon he together with William Embry Gent his Security Enter'd into & Acknowledged their Bond, According to Law, for that Purpose. And the Court do adjudge that he take four Pence ferriages for a Man & Seven Pence half Penny for a Man & Horse & No more.

5 March 1754, Page 578
On the Petition of Sundry Inhabitants of this County It is Ordered that a Bridge be built over Cubb Creek at or near Richard Dudgeons's. And that Cornelius Cargill, Abraham Martin & William Caldwell Gent. let the same to undertakers to be built in such manner upon such terms & Conditions as they shall think most [blank in book] & Convenient.

7 May 1754, Page 3
George Elliot is appointed Surveyor of the Road whereof John Twitty was late Surveyor, And it is Ordered that he together with the following Assistance, to wit, The male labouring Tithable Persons belonging to the said Elliot, Joseph Boswel and his those belonging to Thomas Pettus, John Chandler, William Lax Joseph Davis Philip Cockerham, William Stones & his Male labourers, do forthwith Clear & keep the same in Repare According to Law.

7 May 1754, Page 3
Edward Goode is appointed Surveyor of that Part of the road Whereof Richard Witton Gent was late Surveyor, leading from Ellidges road to Twittys Road, And it is ordered that he together with the following Assistance, to wit, Jeremiah Hatcher, Nicholas Hobsom, John Hankins, George Platt and Philip Poindextor with their Male Labouring Tithable Persons do forthwith Clear & Keep the same in Repair According to Law.

7 May 1754, Page 4
Richard Witton Gent is Continued Surveyor of of that Part of the Road Whereof he was late Surveyor, leading from the fork of his Road to the fork of Ellidges Road, And it is Ordered that he together with his own male labouring Tithable Persons do forthwith Clear & keep the same in Repair According to Law.

7 May 1754, Page 4
On the Petition of Peter Fontaine Gent for a Road to be laid off & Clear'd the best & most Convenient way from Willinghams Road to Reedy Creek Church, It is Ordered that John Bacon Drury Allen and Daniel Claiborne (or any two of them being first Sworn &c) do View & Examine the way where such Road is Petition'd for & Report to the Next Court the Conveniency or inconveniency thereof.

7 May 1754, Page 4
On the Motion of John Twitty for a Road to be laid off and Clear'd the best and most Convenient way from a Place call'd Tusling Quarter in this County to the Road called Elledges Road, It is Ordered that Francis Bracy, Martin Fifer & Anthony Hughes or any two of them (being first Sworn &c) do View & Examine the Way Such Road is Petition'd for, and Report to the next Court the Conveniency, or inconveniency thereof

7 May 1754, Page 5
On the Petition of David Logan & others, for a Road to be laid off and Clear'd, the best and most Convenient Way from Stewards Ferry on Stanton to Joseph Mortons Quarter on the said river, It is Ordered that John Middleton, Thomas Stewart & John Logan or any two of them (being first Sworn & c) do diligently View & Examine the way where such Road is Petition'd for, and Report to the Next Court the Conveniency &c.

7 May 1754, Page 5
On the Petition of Hugh Boston, Setting forth that there is a Certain Public Road in this County which leads through or Near his Plaintation, Praying leave to turn the said Road further from his said Plantation, David Garland Gent is Appointed and desired to View & Examine the Way where such Road is intended to be turned, and Report to the next Court the Conveniency or inconveniency thereof.

7 May 1754, Page 5
John Jennings is appointed Surveyor of the Road leading from Bairs Ellement Creek, to the Road Called Dyers Road, whereof John Williams was late Surveyor, And it is Ordered that he together with all the assistance that assisted the said Williams on the said Road do forthwith Clear & keep the same in Repair According to Law.

7 May 1754, Page 6
William Nance is appointed Surveyor of the Road leading from Maherrin River to the Horsepen branch of Juniper Creek, whereof Thomas Willingham was late Surveyor, And it is Ordered that he together with the following assistance, to wit, William Nance junior, John Nance Senior, Richard Nance, Thomas Nance, Thomas Willingham, William Willingham, W^m. Willingham junior, James Roberts Senior, James Roberts Junior and Patrick Flin with their Male labouring Tithable Persons do forthwith Clear & keep the same in Repair According to Law.

7 May 1754, Page 6
John Glasscock is appointed Surveyor of the Road leading from the Horsepen fork of Juniper Creek to the Road called Randolphs Road whereof Thomas Willingham was late Surveyor, And it is Ordered that he, together with the following Assistance, to wit, Henry Isbell, Thomas Mitchel, Michael Gill, and Francis More Petty with their Male labouring Tithable Persons do forthwith Clear & keep the same in Repair According to Law.

7 May 1754, Page 6
Richard Tomson is appointed Surveyor of the road Whereof Francis More Petty was late Surveyor, And it is Ordered that he together with the assistance that assisted the said Petty on the said Road do forthwith Clear & keep the same in Repair According to Law.

7 May 1754, Page 7
Thomas Crenshaw is appointed Surveyor of the road call'd Waltons road in the room of George Wells, And it is Ordered that he together with the following assistance, to wit, Cornelius Crenshaw John Willingham, Barnaby Wells, Argil Blaxtone, George Wells & Elijah Wells with their Male labouring Tithable Persons and the Male labouring Tithables Belonging to George Walton at his Mill Plantation, do forthwith Clear & Keep the same in Repare According to Law.

7 May 1754, Page 7
William Petty Pool is Ordered (together with the Assistance Convenient thereto) to lay open & Clear a Road the best & most Convenient way from Willinghams Bridge on ledbetter Creek to Reedy Creek Church, and that he keep the same in Repair According to Law.

7 May 1754, Page 8
Grand Jury Presentments
...the Overseers of the Road leading from Ruffins road to Bryery River ...
* * *
...the Overseer of the road leading from Jeffersons ferry to the Church Road...
* * *
...the overseer of the Road leading from Perrin Aldays to Thomas Vernons junior...
...the Overseers of the Road leading from this Court house to the Horse Pen Creek...

7 May 1754, Page 9
Owen Sullivant is appointed Surveyor of the Road whereof Thomas Jones was late Surveyor, And it is Ordered that he together with the assistance that assisted the said Jones on the said Road, do forthwith Clear & keep the same in Repare According to Law.

7 May 1754, Page 11
William Petty is appointed Surveyor of the Road Whereof James Taylor was late Surveyor, And it is Ordered that he together with the following Assistance, to wit, Richard Williams Lazarus Williams, Benjamin Cockrum, John Strone, James Garland, Nathaniel Garland, John Forrest & Jeffery Murrel with their Male Labouring Tithable Persons, & the male labouring Tithables belonging to Samuel DuVal & those of Henry Embry do forthwith Clear & keep the same in Repair According to Law.

7 May 1754, Page 11

Peter Cox is appointed Surveyor of the Road whereof Daniel Hayse was formerly Surveyor, And it is Ordered that he together with all the male labouring Tithable Persons living from Smiths Quarter up Crooked Creek to where one Rivers formerly lived, do forthwith Clear & keep the same in Repair According to Law.

7 May 1754, Page 11
George Philips is appointed Surveyor of the Road Whereof James Amos was late Surveyor, And it is Ordered that he togethr with the Assistance that assisted the said Amos on the said Road do forthwith Clear & keep the same in Repair According to Law.

7 May 1754, Page 11
William Drew is appointed Surveyor of the road Whereof John Cargill was late Surveyor, And it is ordered that he, together with the Male Labouring Tithable Persons that assisted the said Cargill on the said Road do forthwith Clear & keep the same in Repair According to Law.

7 May 1754, Page 12
David Dortch is appointed Surveyor of the road Whereof Dennis Lark was late Surveyor, And it is ordered that he together with the male labouring tithable Persons that Worked under the said Lark on the said Road, do forthwith Clear & keep the same in Repair According to Law.

7 May 1754, Page 13
Order'd that George Vaughn be Extended Surveyor of the Road Whereof he is Surveyor, to the Mountain Creek, And it is ordered that he together with the Male labouring Tithable Persons Convenient thereto do forthwith Clear & keep the same in Repair According to Law.

7 May 1754, Page 13
Henry Gill is appointed Surveyor of the Road whereof David Garland Gent was late Surveyor, And the said David Garland is appointed & Desired to, Nominate assistance to work under the said Gill, And it is Ordered that he together with such assistance as the said Garland shall appoint, do forthwith Clear & keep the same in Repare According to Law.

7 May 1754, Page 13
Nehemiah Frank is appointed Surveyor of the Road Whereof Abraham Martin Gent was late Surveyor, And it is Ordered that he together with the male labouring Tithable Persons that assisted the said Martin on the said Road, do forthwith Clear & keep the same in Repair According to Law.

7 May 1754, Page 13
Order'd William Gee, Henry Gee & Neel Gee, William M^c.Clehomny, John Granger, Crooked Creek Ragsdale, and the male labouring Tithable Persons at Cockes Quarter, be added to the Road Whereof William Saffold is Surveyor.

7 May 1754, Page 13
On the Petition of John Bacon leave is Granted him to make a Bridle Way of the Old Established Path leading from his house to Cox Mill.

7 May 1754, Page 17
On the Motion of John Fuquay Licence is Granted him to keep a Ferry on Stanton River from his own land to the land of Joseph Mayse at the said Mays's late Ferry landing (he giving Security) Whereupon he together with Thomas Nash & James Hunt his Securitys Entered into & Acknowledged their Bond According to Law for that Purpose, And the Court do adjudge that he take three Pence ferriage for a Man Six Pence for a Man & Horse, three Pence for every Wheel, & Six Pence for a Hogshead of Tobacco, and No more.

8 May 1754, Page 28
Samuel Johnson is appointed Surveyor of the Road whereof Robert Woods was late Surveyor from Little Ronoke bridge to the low Ground on the North side of Wards ford Creek, And it is ordered that he together with the following assistance, to wit, three of the Male labouring Tithable Persons belonging to Clement Read, two of those belonging to Mary Gwinn John Atkins William Atkins, Benjamin Atkins, William Stanfield David Jones, William Johnson, & Hugh Creighton do forthwith Clear & keep the same in Repair According to Law.

8 May 1754, Page 29
Robert Woods is Continued Surveyor of that Part of the Road Whereof he was late Surveyor, leading from the low Ground on the North Side of Wards fork Creek to Coles Road, And it is Ordered that he together with the following Assistance to wit, Osborne Keeling, Wilson Mattox, John Mount, Andrew Cannaday Ensworth Middleton, Daniel Humphris, Richard Berry, George Hannah, John Austin, William Adams & [blank in book] Hays with their Male labouring Tithable Persons do forthwith Clear & keep the same in Lawful Repair.

8 May 1754, Page 29
Daniel May is appointed Surveyor of the Road Whereof Charles Talbot was late Surveyor, And it is Ordered that he, together with the assistance that assisted the said Talbot on the said Road, do forthwith Clear & keep the same in Repair according to Law.

4 June 1754, Page 63
Anthony Hughes and Martin Fifer two of the Persons appointed by an Order of the last Court to View and Examine the way where a Road was Petition'd for by John Twitty leading from Tusling Quarter, to Elledges Road, this day Return'd a Report thereon, And thereupon the same is establishd a Road, And the said John Twitty is appointed Surveyor thereof, And it is Ordered that he together with his Male Labouring Tithable Persons, Anthony Hughes & his, do forthwith lay open Clear and keep the same in Repair According to Law.

4 June 1754, Page 63
Thomas Stewart and John Logan two of the Persons appointed by an Order of the last Court, to View and Examine the Way Where a Road was Petition'd for by David Logan & others, to lead from Stewarts Ferry on Stanton River, to Joseph Mortons Quarter, this day Returnd a Report thereon, And thereupon the same is establishd a Road, And the said Thomas Stewart is appointed Surveyor thereof, And it is Ordered that he together with William Redman and the Male Labouring Tithables Under him, John Worthy & Thomas Handcock do forthwith Clear & keep the same in Repare According to Law.

5 June 1754, Page 75
William Marrable Surveyor of the High Way from this Court house to Marrables Ordinary in this County, having been Presented for not keeping the same in Repair, This day appear'd in Court, and on hearing the arguments thereon both sides it is Considered by the Court that, the said Presentment be Dismis'd.

5 June 1754, Page 76
Mackness Goode the Surveyor of the High Way from Marrables Ordinary to Kings Road in this County, Who Stands Presented by the Grand jury for not keeping the same in Repair, This day appear'd in Court, And on hearing the Arguments thereon on both sides it is Considered that the Said Presentment be Dismis'd.

6 June 1754, Page 110
Order'd that the following male Labouring Tithable Persons, to wit, John Ragsdale Junior and his Negro, John Ragsdale Senior & his, William Wallace & his Son, Edward Cuttilo & his Son Abraham, James Lowman, James Cohoon, Israel Bron & his Negro, Hugh Wyley, James Gordon John Cuttilo & his two Negros, Drury More, James McKenny, those belonging to John Jennings, William Burgamy & his, do Assist John Jennings in Clearing & keeping in Repair the Road, Whereof the said Jennings is Surveyor.

6 June 1754, Page 110
For Reasons appearing to the Court, It is Ordered that William Caldwell & Abraham Martin Gent^l. do View the Place where a bridge hath latly been Ordered to be built over Cub Creek near Richard Dudgeons's, & Report to the Next Court the Conveniency or inconveniency thereof.

7 June 1754, Page 111
On the Petition of Sundry Inhabitants of Cub Creek in this County, Praying a Bridge over a Miry Branch in the low Grounds of the said Cub Creek near the Bridge over the said Creek, It is ordered that William Caldwell & Abraham Martin Gent^l. do View the Place where such Bridge is Petition'd for, and Report to the Next Court the Conveniency thereof &c.

2 July 1754, Page 134
Order'd that a bridge be Errected & built over little Ronoke River at or near a Place Called the Mossing foard, And Clement Read & Thomas Bouldin are appointed and Desired to lett the Same to Undertakers upon such Terms, & to be built in such manner and form, as they in their Discretion shall think fit & Convenient.

4 July 1754, Page 154
Our Sovereign Lord the King Plaintif
    against
    on a Presentment.
The Surveyor of the Road from Jefferson Ferry to the Church Road Defendant
For Reasons Appearing to the Court his Suit is Ordered to be Dismised.

4 July 1754, Page 161
William Williams Surveyor of a high Way in this County having been Presented for not keeping in Repair a Bridge over cocks Creek in this County, And not Appearing to make his defence, it is Considered that he make his fine with our Sovereign Lord the King by the Payment of

fifteen Shillings Current Money, to his Majestys use And that he Pay the Costs of these Proceedings, And it is Ordered that he be taken &c.

4 July 1754, Page 161
Our Sovereign Lord the King ------ Plaintiff
     Against
     On a Presentment
The Surveyor of the Road from Twittys to Jennings's-Defendant
Leonard Claiborne junior Attorney for our Sovereign Lord the King comes into court, and saith that he will not further Prosecute against the Defendant on the Presentment aforesaid. And thereupon the same is Ordered to be Dismised.

4 July 1754, Page 162
Our Sovereign Lord the King ------ Plaintiff
     against
     On a Presentment
The Surveyor of Stokes's Road ---- Defendant
Leonard Claibourne Junior Attorney for our Sovereign Lord the King comes into Court, and saith that he will not further Prosecute against the Defendant on the Presentment aforesaid. And thereupon the same is Ordered to be Dismis'd,

7 August 1754, Page 173
Mathew Marable is Appointed Surveyor of the Road Leading from this Court House to the said Marables, and it is Ordered that he together with the Male Labouring tithable Persons Convenient thereto forthwith Clear and keep the same in Repair According to Law.

7 August 1754, Page 173
William Edwards is Appointed Surveyor of the Road Leading from Little Bair Creek to Falling River, And it is Ordered that David & William Caldwell do divide & Proportion the Hands Convenient thereto between the said Edwards & the Adjasant Surveyors, And it is also Ordered that the said Edwards with his Hands aforesaid forthwith Clear & keep the said Road in Repair According to Law.

1 October 1754, Page 191
On the Motion of George Holloway, leave is Granted him to Turn Jeffersons Ferry Road into the Church Road.

1 October 1754, Page 192
George Holloway is Appointed Surveyor of the Road Call'd Jeffersons Ferry Road and of the Road Leading from Miles's Creek to Allen's Creek and it is Ordered that he together with the male labouring Tithables Convenient thereto forthwith Clear & keep the same in Repair According to Law.

5 November 1754, Page 213
On the Petition of the Inhabitants of Cubb Creek in this County for a Road from Randolphs Roaling Road the nearest and best way to the Ridge Road that goes through Prince Edward County. It is Ordered that Thomas Harvy and Christopher Parsons (they being first Sworn before a Magistrate of this County) View the Way thereof, and make a Report to the next Court.

5 November 1754, Page 213
Henry May is Continued Surveyor of the Kings Road from Hudsons Foard to the Court House Road and Mackerness Goode from thence to Randolphs Road, and it is Ordered that all the Male Labouring Tithables Convenient thereto Assist them in Clearing and Keeping the said Roads in Repair.

5 November 1754, Page 213
On the Petition of Samuel David for Leave to turn the Road that goes over Wards Fork bridge, It is Ordered that David Caldwell and Robert Woods (they being first Sworn) do View the same and Return a Report thereof to the next Court.

5 November 1754, Page 213
William Almond is Appointed Surveyor of the Road leading from Prince Edward County line to the little Bair Creek, and it is Ordered that he together with all the Male Labouring Tithables Convenient thereto do forthwith Clear and keep the same in Repair According to Law.

5 November 1754, Page 213
Richard Womack is Appointed Surveyor of the Road leading from Thomas Vernons to the Head of Bryery River, and it is Ordered that the said

Womack together with the male Labouring Tithables Convenient thereto do forthwith Clear and keep the same in Repair According to Law.

6 November 1754, Page 226
Abraham Martin and William Caldwell Gent[l]. who were formerly Appointed by this Court to view the Place where a Bridge was formerly Order'd to be Built Over Cubb Creek, this day Return'd a Report that the same is not Necessary nor Convenient, Therefore the former Order is Ordered to be Canceld &c

7 November 1754, Page 235
County Levy

To Samuel Holms for setting up sign Boards ...
    20
To Thomas Jones for setting up three sign Boards ...
    75
To George Holloway for setting up four sign Boards ...
    48
To James Coleman for setting up Six sign Boards ...
    72

* * *

To Richard Ship for Repairing a Bridge Over Little Ronoak ...
    150

* * *

To Richard Witton and Mathew Marable Gentl. for a Bridge by John Cox's ... L29.19.0

* * *

To Thomas Bouldin & Clement Read for a Bridge over Little Ronoak ...
    L15.19.0

7 November 1754, Page 238
Ordered that Lyddal Bacon Gent[l]. do Address Amelia Court to Appoint Persons to Let a Bridge Over Notoway River at Hampton Wades, and that the said Bacon together with the Person or persons that shall be Appointed by the said Court to let the same to Undertakers to Build the same.

8 November 1754, Page 243
Ordered that William McCadoe and Jeremiah Hatcher with their male Labouring Tithables be added to the Road whereof Edward Good is Surveyor, and John Hawkins is Appointed Surveyor of the Road whereof the said Good is Surveyor

8 November 1754, Page 243
On the Motion of Pinkethman Hawkins for Leave to Cut a Bridle Way from his House to Mitchells Road. It is Ordered that Anthony Hughes and William Twitty (the being first Sworn &) do View the best & Convenentest Way thereof and make a Report thereof to the next Court

4 February 1755, Page 258
Ordered that a Bridge be built over a small Branch on the Lower side of Cubb Creek near the Bridge over the said Cubb Creek and it is Ordered that William Caldwell and David Caldwell treat with Workman to Build the same on such Terms in such Manner and for what Consideration they shall think Proper --

4 February 1755, Page 261
On the Motion of James Hunt Gent^l. Licence is Granted him to keep a Ferry in this County on Stanton River from his Own Land to the Land of one Abraham Abney, he giving Security, Whereupon he together with Thomas Nast Gent^l. his Security entered into and Acknowledged their Bond According to Law for that Purpose, And the Court do Adjudge that he shall Seven pence half penny ferriage for a Man & Horse and three pence three farthings for a Man and three pence for a Wheel and no more.

5 February 1755, Page 269
Lyddal Bacon and William Embry Gent^l. are Appointed and Desired to View and Value the Bridge Built by Richard Stokes over Meherrin River, and it is Ordered that they make a Report of the True Value thereof here to the next Court.

5 February 1755, Page 272
On the Motion of Julius Nichols for Leave to Turn the Road leading from Brunsick line to his Ferry in this County, It is Ordered that the said Julius, William Riddle and Richard Fox or any two of them, they being first Sworn Before a Justice of this County do View the most convenient way thereof and Report thereon to the next Court to be held for this County

7 February 1755, Page 289
On the Motion of James Roberts to have a Road Clear'd from Willinghams Road to his Mill, It is thereupon Ordered that Daniel Claiborne, William Pool, Thomas Dupry (or any two of them) they being first Sworn before a Justice of this County, do view the best way thereof and make a Report thereof to the next Court.

7 February 1755, Page 289
On the Motion of James Roberts, leave is given him to Clear a Bridle Way from his Mill to Waltons Court House Road, and it is Ordered that Christopher Sneed and Thomas Williamson do Direct and lay of the same the most Convenient Way thereof

4 March 1755, Page 293
On the Motion of William Royster Licence is Granted him to keep a Ferry in this County on Ronoke River from his Own Land to the Land of William Byrd Esqr., on his giving Bond & Security at the next Court for that Purpose, And the Court do Adjudge that he shall take Seven Pence half Penny ferriage for a Man & Horse and three Pence three farthings for a Man and no more.

4 March 1755, Page 294
On the Motion of Henry Isbell for leave to Turn the Road near his House. It is Ordered that that James Taylor, Thomas Bedford and Charles Sullivent, they being first Sworn before a Justice of this County, do View the best Way thereof and make a Report of their View to the next Court

1 April 1755, Page 299
Mark Thornton is Appointed Surveyor of the Road Whereof Daniel Malone was late Surveyor, And it is Ordered that the together with the Assistance that Assisted the said Malone on the said Road do forthwith Clear and keep the same in Repair According to Law.

1 April 1755, Page 300
Ordered that the Old Road leading from Wisons Road to Hatchers Road be laid open & Cleared, and George Elliot is Appointed Surveyor thereof, and it is Ordered that he together with following assistance, to wit, Thomas Pettis's Male Labourings Tithables Joseph Bozwells Male

Labouring Tithables, the said Elliots Male Labouring Tithables, William Stones Male Labouring Tithables and Joseph Davis's male Labouring Tithables do forthwith Clear and keep the same in Repair According to Law.

1 April 1755, Page 301
William Royster together with Pinkethman Hawkins his Security came into Court, and Entered into and Acknowledged their Bond for a Ferry which the said William Obtained a Licence for at the Last Court

1 April 1755, Page 302
William Watkins is Appointed Surveyor of the Road leading from Bryery River to George Moores, and it is Ordered that he, together with the following Assistance, to wit, his own, George Moore's John Coles, Richard Hills and David Ealbanks Male labouring Tithable Persons do forthwith Clear & keep the same in Repair According to Law.

1 April 1755, Page 304
Julius Nichols & William Riddle who where formerly Appointed to View the Road leading from Brunsick County line to the said Julius's Ferry, this day Return'd a Report that it will be Convenient to Turn the same, and Julius Nichols is appointed Surveyor of the said Road, and it is Ordered that he together with the Assistance that that Assisted the former Surveyor thereof do forthwith Clear & keep the same in Repair According to Law.

4 April 1755, Page 333
Silvanus Walker Appearing on being Summon'd to shew Cause why he did not keep in Repair the Bridge by him built over Maherrin River & for Reasons Appearing it is Ordered to be Dismised.

4 April 1755, Page 334
On the Motion of James Hunt Gent[l]. for a Road to be laid off and Cleared the best and most Convenient Way from Mays's Road to his Ferry, It is Ordered that Thomas Spraggons, William Spraggons and Joseph Bayse (being first Sworn &c) do View & Examine the Way where such Road is Petitioned for and Report to the next court the Conveniency or inconveniency thereof.

6 May 1755, Page 337
On the Petition of Thomas Lowry, setting forth that there is a Certain Public Road Called Flat Rock Road which leads through his Plantation Praying Leave to Turn the said Road, It is thereupon Ordered that John Ragsdale, William Wallis And William Berry (they being first sworn &c) do View & Examine the Way where such Road is intended to be Turn'd and Report to the next court the Conveniency or inconveniency thereof.

6 May 1755, Page 337
David Garland Gent is Appointed Surveyor of the Road leading from the ford over Meherrin River at Francis Rays, to Captn. John Jennings Ordinary, and it is Ordered that he together with all the male Labouring Tithable Persons Convenient thereto, do forthwith Clear & keep the Same in Repair According to Law.

6 May 1755, Page 339
James Thomson Barden is Appointed Surveyor of the Road whereof Isaac Hudson was late Surveyor, and it is Ordered that he together with the Assistance that Assisted the said Hudson do forth with Clear & keep the same in Repair According to Law.

7 May 1755, Page 348
Grand Jury Presentments
...the Overseer of the Road from the Mossing ford Bridge to Captain Thomas Bouldins, for not keeping the same in Repair ...
* * *
...the Overseer of the Road from Capt Thomas Boudlins to Ash Camp for not keeping the same in Repair, and against the Surveyor of the Road from Ash Camp to Meherin, a Presentment against the overseer of the Road from Mitchells Ferry to this Court House for not keeping the same in Repair, a Presentment against the Overseer of the Road called Wittons Road from the fork this side of Richard Wittons to the fork below the North Meherin, and a Presentment against the Overseer of the Road from Saffolds ford on Meherin River to the Road at Capt John Jennings's ...

7 May 1755, Page 349
Spettle Pully is Appointed Surveyor of the Road whereof George Holloway was late Surveyor, and it is Ordered that the Assistance that Assisted the said Holloway do Assist the said Pully in clearing and keeping the same in Repair According to Law.

7 May 1755, Page 349
George Holloway is Appointed Surveyor of the Road leading from Ruffins old mill to the Church Road, and it is ordered that he together with Ruffin's and John Speeds male Labouring Tithable Persons do forthwith Clear and keep the same in Repair According to Law.

7 May 1755, Page 349
Benjamin Farmer is Appointed Surveyor of the Road leading from Stewarts Ferry to the Mossing Ford, whereof Daniel May was late Surveyor, and it is Ordered that he with Assistance that Assisted the said May, do forthwith Clear and keep the same in Repair According to Law.

7 May 1755, Page 349
Jonathan Vernon is Appointed Surveyor of the Road leading from the fork of Dunavant Creek to the Little Ronoke uper Bridge and it is Ordered that he together with all the male labouring Tithables Persons Convenient thereto do forthwith Clear & keep the same in Repair According to Law.

7 May 1755, Page 353
Michael Mackie being Summon'd to Appear to shew cause by he did not keep in Repair a Bridge by him Built over Maherrin River this day Appear'd, and it is Ordered that the said Summons be dismis'd and that the said Michael pay the Costs in this Behalf Expended.

7 May 1755, Page 353
Thomas Nash and Thomas Bouldin Gent^l. are Appointed let a Bridge to be Built over Little Ronoake River at the Place where the uper Bridge formerly stood over the said River, to undertakers in such manner & on such terms as they shall think fit.

3 June 1755, Page 359
Ordered that Joseph Morton, Stephen Collins, William Standfield and John Pratt, or any three of them (they being first sworn before a Justice of this County) do View & Examine the nearest & Best Way for a Road from the Bridge Over Little Ronoak River above Clement Reads to Coles Road and Report to the next Court the Conveniency or inconveniency thereof.

4 June 1755, Page 365
Owen Sullivant a Surveyor of High Way in this County, who stands Presented by the Grand Jury, for not keeping the same in Repair, Appeared and on hearing the Arguments on both sides, it is Considered that the said Presentment be Dismis'd.

4 June 1755, Page 365
Richard Hix a Surveyor of a High Way in this County, who stands presented by the Grand Jury, for not keeping the same in Repair, Appeared and on hearing the Arguments on both sides, it is Considered that the the said Presentment be Dismis'd.

4 June 1755, Page 365
Philip Jones a Surveyor of a High Way in this County, who Stands Presented by the Grand Jury for not keeping the same in Repair, Appeared and on hearing the Arguments on both sides, It is Considered that the said Presentment be Dismis'd.

4 June 1755, Page 365
John Hawkins a Surveyor of a High Way in this County, who Stands Presented by the Grand Jury, for not keeping the same in Repair Appeared and on hearing the Arguments on both sides, It is Considered that the said Presentment be Dismis'd.

4 June 1755, Page 365
Samuel Comer is Appointed Surveyor of the Road whereof Richard Hix was late Surveyor, and it is Ordered that he, together with Assistance that Assisted the said Hix, do forthwith Clear and keep the same in Repair According to Law.

4 June 1755, Page 368
Ordered that Thomas Hall Moses Hall, David Roberts, Francis Worsham Francis Petty, Michael Gill, Thomas Mitchell, Richard Jones, Richard Hix, George Foster & his Male Labouring Tithable Persons do Assist Philip Jones & clearing & keeping in Repair the Road whereof he is Surveyor.

4 June 1755, Page 369
Ordered that Thomas Jones together with the Hands at Clement Reads uper Quarter do Clear the Road from the Mossing Ford Bridge into Randolps Road.

4 June 1755, Page 376
Richard Womack a Surveyor of a High Way in this County who stands Presented by the Grand Jury for not keeping the same in Repair, appeared and on hearing the Arguments on both sides It is Considered that the said Presentment be Dismis'd.

7 October 1755, Page 3
Henry Decker is Appointed Surveyor of the Road leading from Davis's ford to Claunch's Ordinary in Room of James Coleman, & It is Ordered that the Decker, with the Assistance that assisted the said Coleman on the Said Road, do forth with Clear & keep the same in Repair According to Law.

7 October 1755, Page 3
John Glass is appointed Surveyor of the Road leading from Mitchells Ferry to Claunch's Ordinary, in the Room of James Coleman, and it is Ordered that the said Glass with the Assistance that assisted the said Coleman, do forthwith Clear & keep the same in Repair According to Law.

7 October 1755, Page 4
John Ragsdale, William Wallice, & William Barry, who were formerly Appointed to View the Road leading thro' Thomas Lowrys Plaintation this day Returned their Report, that It will be Convenient to Turn the said Road, which is Ordered to be Recorded.

4 November 1755, Page 33
Thomas Addams is Appointed Surveyor of the Road whereof John Thompson was late Surveyor, and it is Ordered the said Addams, with the

Assistance that assisted the said Thompson, on the same Road, do forthwith Clear & keep the same in Repair According to Law.

4 November 1755, Page 34
Grand Jury Presentments
... the Overseer of the Road from this Court House to M^r. Marables for not keeping the Road in Repair ...

5 November 1755, Page 36
On the Petition of Sundry Inhabitants of this County, Ordered that a Bridle Way be laid off and Clear'd the best & most Convenientest Way from the County line near William Watsons to Cubb Creek Road near John Wallers, and Alexander Joyce and Richard Austin are Appointed Surveyors thereof and it is ordered that the said Joyce & Austin, together with all the Male Labouring Tithables Convenient thereto do forthwith lay open & Clear the same

6 November 1755, Page 44
Ordered that the following Persons be added to the Road whereof Samuel Johnson is Surveyor, to Assist the said Johnson in Clearing & keeping the Same Repaired, to wit, John Pleasants & Richard Wards Male Labouring Tithables, James Nelson, George Harris & John Butler

7 November 1755, Page 58
County Levy
To Samuel Perrin for Building a Bridge ...
                L6.9.6

* * *

To James Hunt Assignee of Jacob Robinson for Building a Bridge Over Cubb Creek ...

                                            9.2.6

3 December 1755, Page 69
Mathew Marable Surveyor of a High Way in this County, who stands Presented by the Grand Jury, for not keeping the same in Repair, Appeared and on hearing the Arguments on both sides, It is Considered that the Said Presentment be Dismisd

3 December 1755, Page 72
Stephen Evens is Appointed Surveyor of the Road, whereof Mathew Marable was late Surveyor, and, it is Ordered that the said Evens together with the hands that Assisted the said Marable do forthwith Clear & keep the same in Repair According to Law.

4 February 1756, Page 88
On the motion of Mathew Marable Leave is granted him to Turn the Road near his House to Turn out above the said Marables where McNess Goodes Path Crosses fence along the said Path to a Ridge that goes the back of John Hights from thence down the same the best and most Convenient way to the Road below the said Marables and it is Ordered that the said Marable do Lay Open and Clear the same with his Own hands, and if it shall appear that the way that he Intends to Clear the Road, is not Convenient, after a View thereof, that then the Road is to Continue where it is now

4 February 1756, Page 90
James Parrott is Appointed Surveyor of the Road whereof Henry Decker was late Surveyor, leading from Claunches Ordinary to Kings ford -- and it is Ordered that the said James together with the Assistance that Assisted the said Henry Decker on the said Road, do forthwith Clear and keep the same in Repair According to Law.

1 June 1756, Page 114
Stephen Mallet is appointed Surveyor of the road leading from Delonys Old Ordinary to Cocks Creeks, and it is Ordered that he together with all the Male labouring Tithables Convenient thereto do forthwith Clear and keep the same in Lawfull Repair.

1 June 1756, Page 116
On the Petition of Sundry Inhabitants of this County, Setting forth that the Inhabitants on the South Side of Ronoke River have for a Long time labour'd under a great Inconveniency for want of a Ferry over the said River to go to and from Church, Court and Market, Praying this Court would Erect a Ferry at some Convenient Place for that Purpose, It is thereupon Ordered that if William Royster will Set Over the said River all or any of the Inhabitants of this County that Lives on the South Side of the said River to and from this Court House, to and from any Church, & to and from Market for the Quantity of Five hundred Pounds of Tobacco P annum, that a Ferry for that Purpose be Erected at his Landing, and if he well not, that then the Same be Erected at the Landing of George Curry upon the same Conditions

1 June 1756, Page 116
Ordered that Henry May Surveyor of the Road leading from Fontains Ferry in this County do Clear the Road from the Ferry Round the Said Fontains Plantation the nearest and Best Way to the Ferry Road.

1 June 1756, Page 117
On the Motion of Richard Witton Gent^l. and for Reasons Appearing, Leave is Granted him to keep Gates at his Plantation a Cross the Road.

2 June 1756, Page 135
On the motion of Joseph Williams Gent^l, and for Reasons Appearing, It is Ordered that Hands on Waltons Road and Johnsons Road be by the said Joseph Williams and Isaac Johnson devided & Proportioned to the Several Surveyors thereof as they shall think fit; and it is further Ordered that the said Johnsons Road be Extended to Waltons Road, and that the said Joseph Williams, Isaac Johnson, Joseph Johnson & Stephen Wood or any three of them (being first Sworn before a Justice of this County) do View & Examine the most Convenient Way thereof and make a Report to the next Court.

6 July 1756, Page 154
On the Petition of Clack Courtney, setting forth, that some time ago he became Security for Edward Carter to John Speed and George Baskervile (trustee Appointed by this Court to Let the Building of a Bridge to undertakers, Over Miles Creek), for the said Edwards Building the said Bride for the Consideration of Seven Pounds Ten Shillings Current Money, -- that the said Edward never finished the said Bridge, and that since the said Clack Courtney hath finished the said Bridge According to the Agreement made between the said John Speed and George Baskervile, and the said Edward Carter, and that he hath delivered it to the Trustees aforesaid; Bring this Court to allow him for the same, Therefore It is Ordered that the late Collector of this County do pay him the said Clack the aforesaid Sum of Seven Pounds Ten Shillings out of the Depositum in his Hands after deducting thereout two hundred & twenty Pounds of Nett Tobacco & Thirty Shillings. The Costs of the Suit lately determined in this Court between the said John Speed & George Baskervile Complainants and the Said Edward Carter Defendant.

6 July 1756, Page 155
William Riddle is Appointed Surveyor of the Road Called Julius Nichols's Road leading from Humphry Heweys to the Lower End of the said Road, and it is Ordered that he, together with all the Male Labouring Tithable Persons Convenient thereto, do forthwith Clear & keep the same in Repair According to Law.

6 July 1756, Page 155
Jacob Bugg is Appointed Surveyor of the Road leading from Allens Creek to Miles Creek in the Room of Spettle Pully and It is Ordered that he together with the Hands that Assisted the said Pully, do forthwith Clear and keep the same in Repair According to Law.

6 July 1756, Page 155
George Baskervile is Appointed Surveyor of the Road leading from Jeffersons Ferry to the Church Road, and it is Ordered that he together will all the Male labouring Tithable Persons Convenient thereto do forthwith Clear & keep the same in Repair According to Law.

6 July 1756, Page 156
William Wallice is Appointed Surveyor of the Road leading from Bears Ellement to the Fork of Dyers Road, and it is Ordered that he, together with all the Male Labouring Tithables Convenient thereto do forthwith Clear and keep the same in Repair According to Law.

3 August 1756, Page 171
On the Petition of Sundry Inhabitants of the uper End of this County, Setting forth that they labour under a great Inconveniency for want of a Road from Francis Grahams ford on Cubb Creek to to Wards Fork Bridge Commonly Called David Caldwells Road, and that the Way hath been View'd and found to be Convenient, It is thereupon Ordered James Rutherford together with all the male Labouring Tithables Convenient to forthwith lay open & keep the same in Repairs According to Law.

3 August 1756, Page 176
Ordered that John George Pennington William Pinnin, and John Ezell, being first Sworn, do View & Examine the Road Leading from Ingrams Road to Mize's Ford Road, and that they do Report to the next Court the Conveniency or inconveniency thereof.

3 August 1756, Page 178
Ordered that Richard Palmer, John Bracy, Michael McNeil, and Daniel McNeil or any three of them, being first Sworn, do View & Examine the Way From Mitchells Ford by Daniel Mitchells & Austin Spears's to the Country Line, and Report to the next Court the Conveniency or inconveniency thereof.

3 August 1756, Page 178
Ordered that Samuel Maning Continue the Road whereof he is Surveyor to Miles Creek Bridge and that he & the hands now under him do forthwith lay open & keep the same in Repair According to Law.

4 August 1756, Page 184
On the motion of Thomas Anderson, for a Licence to keep a Ferry Over Roanoak River Opposite to Mitchells Ferry. For Reasons Appearing to the Court is Ordered to be Rejected

4 August 1756, Page 185
On the Petition of Sundry Inhabitants of this County for a Road to be laid off and Clear'd, the best and most Convenient Way from Cargills Road to be Open'd out of Cargills Road (at the Tob of the Hill Opposite Cornelius Cargills to Willinghams Bridge on Maherrin River, It is Ordered that Memucan Hunt, John Cargill, Stephen Evens, Richard Scruggs and McNess Goode or any three of them (being first Sworn &ca) do diligently View & Examine the Way where such Road is Petition'd and Report to the next Court the Conveniency or inconveniency thereof.

7 September 1756, Page 187
John George Pennington, William Pinnell and John Ezell the Persons Appointed at the last Court to view & Examine the Road leading from Ingrams Road to Mizes Ford Road, this day Returned a Report thereon, and thereupon the Same is Established a Road, and the said John George Pennington is Appointed Surveyor thereof, and It is Ordered that he together with John Mize, James Edmonds, John Ezell, John Bates, Henry Bates George Ezell, Michael Ezell William Pinner, Robert Connell William Tucker and George Chavus do forthwith clear and keep the same in Repair According to Law.

7 September 1756, Page 188
Richard Palmer and others who were Appointed by an Order of the last Court to View & Examine the Way where a Road was Petition'd for by

Sundry Inhabitants of this County to lead from Mitchells Ford by Daniel Mitchells and Austin Spearss to the Country Line, this day Return'd a Report thereon and thereupon the same is Established a Road, and Francis Wagstaff is Appointed Surveyor thereof, and it is Ordered that he together with his Own Male Labouring Tithables, Theophilus Field's, John Bracy's, Daniel Mitchell and Austin Spears do forthwith lay open Clear and keep the same in Repair According to Law

8 September 1756, Page 200
On the Petition of George Walton, for a Road to be laid Off Clear'd the best and most Convenient Way from his dwelling House to Little Roanoak Church, It is Ordered that William Gill and Michael Gill, being first Sworn, do View & Examine the Way where such Road is Petition'd for, and Report to the next Court the Conveniency or inconveniency thereof.

8 September 1756, Page 200
Ordered that the Sherif Summon the Several Surveyors of the Road leading from this Court House to Randolphs Road, to Appear here at the next Court, to shew cause why they do not keep the same in Repair.

8 September 1756, Page 200
William Soffold Surveyor of the High Way in this County, having been Presented for not keeping in Repair the Road whereof he is Surveyor and not Appearing to make his defence, It is Considered that he make his fine with our Sovereign Lord the King, by the Payment of fifteen Shillings Current Money to his Majestys Use, and that pay the Cost of these Proceedings, and it is Ordered that he be taken &c[a].

8 September 1756, Page 201
On the Petition of John Hobson & others, for a Road to be laid off and clear'd, the best and most Convenient way from Ready Creek Church to the Forks of Wittons Road, It is Ordered that John Bacon, David Gentry & Simon Gentry (being first Sworn &c[a].) do diligently View & Examine the way where such Road is Petition'd for, and Report to the next Court the Conveniency or inconveniency thereof.

5 October 1756, Page 214
Michael Gill and William Gill who were Appointed by an Order of the Last Court, to View & Examine the Way Petition for Road by George Walton, from the said Waltons dwelling House to Little Roanoak Church,

this day Return'd a Report thereof, and thereupon the same is Established a Road.

5 October 1756, Page 214
John Bacon Sen[r], Simon Gentry and David Gentry, the Persons Appointed by an Order of the last Court, to View & Examine the Way Petition'd for a Road by John Hobson & others from Reedy Creek Church to the Forks of Col[o] Wittons Road, this day Return'd a Report thereon, and thereupon the same Establish'd a Road.

6 October 1756, Page 226
Ordered that Mathew Marables male Labouring Tithables, William Marables male Labouring Tithables, Branch Tanners Male Labouring Tithables & Joseph Hunt do Assist Stephen Evens in Clearing & keeping in Repair the Road whereof he is Surveyor.

6 October 1756, Page 227
Ordered that Mathew Marable do Clear & Grub the Road by him turn'd out of the Road leading thro' his Plantation According to Law, and that he do the same Immediately

2 November 1756, Page 228
Thomas Spraggons, Williams Spraggons & Joseph Bays, three of the Persons Appointed by a former Order of this Court, to View & Examine the Way Petition'd for a Road by James Hunt Gent[l]. from the said Hunts Ferry to Mayse's Ferry Road, this day Return'd a Report thereon, and thereupon the same is Established a Road, and Charles Hunt is Appointed Surveyor thereof, and it is Ordered he, together with Richard Bookers Male Labouring Tithables John Fuquays Male Labouring Tithables, Joseph Bays, William Mayse, Dennit Abney, Mattox Mayse's Male Labouring Tithables James Hunts Male Labouring Tithables and James Cunningham do forthwith lay open, clear & keep the same in Repair According to Law.

2 November 1756, Page 229
Grand Jury Presentments
... against Feilding Jefferson for not keeping his Ferry Boat in Repair ... ... against the

Surveyors of a Road from the Rev^d M^r. Keys Plantation to Randolphs Road ...

2 November 1756, Page 229
Jacob Robinson Senr is Appointed Surveyor of the Road leading from Hunts Ferry Road to Cubb Creek Bridge, whereof James Hunt Gent^l was late Surveyor, and it is Ordered that he together with William East Senr. John Dickerson Thomas Baughan, John Kiersey Senr. Thomas Kersy, John Keirsey jun^r. William Roberson, John Roberson, John Butler, William Black, David George, William Hardwitch, Robert Andrew, Andrew Cunningham, William Cunningham & Peter Franklin do forthwith Clear & keep the same in Repair According to Law.

2 November 1756, Page 229
On the motion of James Hunt Gent^l. for Leave to Clear a Road from his Ferry to Bedford County Line, It is thereupon Ordered that Thomas Watkins, Michael Prewit & Richard Prewit (being first Sworn) View & Examine the Way where Such Road is Intended to be Cleared, and Report to the next Court the conveniency &c^a.

3 November 1756, Page 238
Ordered that a Bridge be built over Stokes's Mill Creek, at the Place where a Bridge was formerly built in this County, that Lyddal Bacon Gent^l. treat with Workmen to Build the same, in what Manner & for what Consideration he shall think Proper, and that the undertaker do give Bond with Sufficient Security to the said Lyddal Bacon, to Warrant and Maintain the same Seven Years.

4 November 1756, Page 251
Ordered that Mathew Marable Clear & Grub the Road Clear'd by him round his Plantation, According to Law, or that he do Lay open the Way that the Road was formerly Cleared, & that he do the same Instantly.

7 December 1756, Page 259
Nehemiah Frank Surveyor of a High way in this County, having Presented for not keeping the Road in Repair leading from the Horspen Creek to William Keys Plantation, appearing, and on hearing his Excuses, It is Considered and Accordingly Ordered that he make his fine with our Sovereign Lord the King, by the Payment of fifteen Shillings Current Money, to his Majestys use, and that he pay the Cost of these Proceedings, And it is Ordered that he be taken &c^a.

1 March 1757, Page 266
Thomas Harvy and Christopher Parsons, the Persons formerly Appointed by an Order of this Court, to View & Examine the Way Petition'd for Road by Sundry Inhabitants of this County from Randolphs Rolling Road to Prince Edward, This day Return'd a Report thereon, which is Ordered to be Recorded and the said Thomas Harvy is Appointed Overseer thereof, and it is Ordered that he, together with the male Labouring Tithable Persons Convenient thereto, do forthwith Clear & keep the same in Repair according to Law.

1 March 1757, Page 270
Ordered that David George and Andrew Cunningham do Settle & Proportion the Hands to the Several Surveyors of the Road between Little Ronoak Bridge & falling River, as the shall think Proper.

1 March 1757, Page 270
Ordered that Joseph Moreton and Alexander Joyce do Settle & Proportion such Hands to Assist the Surveyor of the Road leading from Little Roanoak Bridge to the County, as they shall think Proper.

5 April 1757, Page 276
On the Petition of John Jones, setting forth that the Plantation whereon he now lives is the Plantation that one Michael Waldrope formerly lived thro which a Public Road passes; that during the time the said Waldrope lived on the said Plantation it was upon his Request, by the Overseer & Clearers of the said Road, agreed that that part which ran thro' his tended Ground (and was of course Prejudice to him) should be turn'd round his Fence provided the said Waldrope would do it himself, In Consequence or which, the Said Waldrope did according open the said Road, & it has been Commonly used. But as there has not been any Order of Court Obtain'd for the Regular turning of the said Road; some People have, merely out of Wantoness, thrown down his Fence, & he fears may Continue so to do, even when his Crop is gone, to his very great detriment Humbly praying that this Court would Establish the way turn'd Round his Fence a Public Way, and on Consideration thereof, the same is Established a Public Road instead of the Road that Ran thro' the aforesaid Plantation

5 April 1757, Page 280
Ordered that Thomas Covington, Henry Isbell & John Glascock (being first Sworn before a Majistrate of this County) do View & Examine the

Way Round the Head of Ash Camp, and see if it be not a more Convenient Way than the Present over Ash Camp, and make Report thereof to the next Court.

5 April 1757, Page 281
Ordered that George Wells and John Foster, (being first Sworn before a Justice of this County, do View and Examine, the Way for a Road from Owles Creek to Cross Maherrin to Mores below the old Road, and Report to the next Court the Conveniency or inconveniency thereof.

5 April 1757, Page 281
Ordered that a Road be laid Open and Cleared the nearest, best, and most Convenientest Way from the Stony Hill by Cornelius Cargill to Kings Road, and John Cargill is Appointed Surveyor thereof, and it is Ordered that he together with the male Labouring Tithables under William Drew do forthwith Lay open, Clear & keep the same in Repair According to Law.

3 May 1757, Page 287
Ordered that the Bridge Over Cubb Creek, be Repaired and that James Hunt and William Caldwell Gent^l. treat with Workmen to Repair the same on such Terms, in Such manner and for what Consideration they shall think Proper, and that the Undertaker do Warrant & Maintain the same Five Years.

3 May 1757, Page 287
Henry Childress is Appointed Surveyor of the Road whereof Benjamin Farmer was late Surveyor, and it is Ordered that he together with the Assistance that Assisted the said Farmer, do forthwith Clear & keep the same in Repair According to Law.

3 May 1757, Page 288
Grand Jury Presentments
...against Thomas Anderson for making Fence a Cross the Road from Col^o Wittons to Mitchells Ferry, Michael Satterwhite, Thomas Easly, Edward Colbreath & John Glass Witness's ...
... against the Surveyor of the Road from Kings to Randolphs Road Commonly Call'd the Court House Road, for not keeping the same in Repair ...

...against the Surveyor of the Road from Kings Road from Randolphs Road to the Court House Road ...

...against the Surveyor of the Road, Randolphs Road from Kings Road to the Irish Road...

...the Surveyor of the Road round Mathew Marables Plantation and Bridge for not keeping the same in Repair...

...against the Surveyor of the Road Commonly Called Cooks Road to the Court House Road...

3 May 1757, Page 289
On the Petition of William Hunt & others for a Road from Col° Birds Mill to the Wood Pecker Creek, It is Ordered that William Harriss (Woodpecker) John Ragsdale and John Wilborne (being first Sworn &c) do View & Examine the Way Petition'd for a Road, and Report to the next Court the Conveniency or inconveniency thereof.

3 May 1757, Page 290
Thomas Covington, John Glascock and Henry Isbell the Persons Appointed by an Order of the last Court, to View the Way Round the Head of Ash Camp to see if it was not a more Convenient Way than the Present Way Over Ash Camp this day Return'd a Report thereon, which is Ordered to be Recorded; And Samuel Comer is Continued Surveyor of the said Road over Ash Camp, and it is Ordered that he together the Hands that were Appointed by a former Order of this Court Assist him on the said Road, and John Glascock, John Richdal, & Thomas Foster, do forthwith lay open & Clear the Way Round the head of Ash Camp aforesaid, and keep the same in Repair According to Law.

3 May 1757, Page 290
John Lucas is Appointed Surveyor of the Road whereof William Saffold was late Surveyor, and it is Ordered that he, together with the Assistance that Assisted the said Saffold on the said Road, do forthwith Clear and keep the same in Repair according to Law.

3 May 1757, Page 297
Ordered that the Surveyors of the Road Leading over the uper Bridge on Little Ronoak River on each side thereof be Summon'd to appear at the next Court to shew cause why they do not keep the same in Repair.

7 June 1757, Page 298
On the Motion of William Mitchell for Leave to turn a Certain Public Way that Passes by his Plantation, It is Ordered that Anthony Hughes, Thomas Chamberland & Pinkethman Hawkins (being first Sworn) do View & Examine the Way where such Public Way is Intended to be turn'd and Report to the next Court, the Conveniency or inconveniency thereof.

7 June 1757, Page 300
On the Petition of Dennis Lark and others for Leave to Clear a Road from Richard Fox's Landing on Roanoke River into Ingrams Road that goes over Meherrin River, It is Ordered that Richard Fox, Joshua Mabry and Dennis Lark (the being first Sworn &c) do View & Examine the most Convenient Way thereof, and Report to the next Court the Conveniency or inconveniency thereof.

7 June 1757, Page 302
On the Petition of Sundry Inhabitants of this County for a Road to be Cleared from Julius Nichols's Ferry to the County Line, It is Ordered that Benjamin Harrison, William Douglass & Drury Malone (they being first Sworn) View & Examine the Way where Such Road is intended to be Cleared, and Report to the next Court the Conveniency or inconveniency thereof.

7 June 1757, Page 307
Ordered that a new Bridge be built over Meherrin River at Willinghams and George Walton and Lyddal Bacon Gentl. are Appointed to let the same to undertakers on such terms and Conditions as they shall think Proper.

5 July 1757, Page 309
Dennis Lark, Joshua Mabrey and Richard Fox, the Persons Appointed by an Order of the last court to View & Examine the Way for a Road to be Cleared from Fox's Landing on Roanoke River into Ingrams Road that goes over Maherrin River, this day Returned their Report thereon, which is Ordered to be Recorded, and Joshua Mabry is Appointed Surveyor thereof, and it is Ordered that he together with Richard Fox & his four Hands, William Gamblin, Claiton Lambert, John Lambert, James Lambert, Humphrey Hewey, James Sparrow, Gabril Hardin -- Stephen Jones, John Bozeman, William Bartlete, Ephraim Mabry George Landford, John Landford William Cleaton, John Cleaton Edward Epps and his Hand, Joseph Hicks, James Hicks, John Mize John Bates, William Ladd Senr. William Ladd jun^r. Garrard Ladd, Thomas Taylor Senr. & his three Hands, John Taylor Thomas Taylor jun^r. John Segant, Wade Ward, Clack

Courtney and his one hand, Lewis Tanner, Thomas Tanner, Lewis Tanner Joseph Bennit, William Bell, Thomas Bell & Amos Timms do forthwith lay open Clear & keep the same in Repair According to Law

5 July 1757, Page 311
On the Petition of Richard Fox and others for Leave to Clear a Road from Richard Fox's Landing on the South side of Roanoak River to the Country line towards Israel Robertsons Mill, It is Ordered that Richard Fox, William Bacon, and William Davis (being first Sworn &c[a]) do View & Examine the Way where such Road is Intented to be Cleared, and Report to the next Court the Conveniency or inconveniency thereof.

5 July 1757, Page 312
Thomas Anderson having being Presented by the Grand Jury for making a fence a Cross a Certain High Way in this County, Appeared, and on hearing the Arguments thereon, the said Thomas is Discharged

5 July 1757, Page 312
Thomas Anderson and John Grainger, Surveyors of a High Way in County, having been Presented by the Grand Jury, for not keeping in Repair the Road whereof they are Surveyors, having been duly Summond and not Appearing, Therefore it is Considered by the Court, that the said Thomas & John make their fine with our Sovereign Lord the King by the Payment of fifteen Shillings Current Money, to his Majestys use, and it is Ordered that they pay the Costs of these Proceedings, and that they be taken &c[a].

5 July 1757, Page 313
William Lydderdale is Appointed Surveyor of the Road whereof Thomas Anderson and Joseph Granger was late Surveyor, and it is Ordered that the said William together with the hands that assisted the said Thomas & Joseph thereon, do forthwith Clear and keep the same in Repair According to Law.

5 July 1757, Page 313
Richard Womack having been Summond to shew Cause why he doth not keep in Lawfull Repair the Road whereof he is Surveyor, Appeared and on hearing the Arugments &[c]. he is Discharged

5 July 1757, Page 313
Ordered that Thomas Nash and William Watkins do Appoint such hands as they shall think fit to assist Richard Womack in Clearing and keeping in Repair the Road whereof he is Surveyor

5 July 1757, Page 316
Joseph Perrin a Surveyor of a High Way in this County, having been Presented by the Grand Jury for not keeping the Road whereof he is Surveyor in Repair, Appeared and on hearing the Arguments &c thereon, he is Acquitted

2 August 1757, Page 343
John Hix is Appointed Surveyor of the Road leading from Col^o. Wittons Road to Reedy Creek Church, and it is Ordered that he together with M^{rs}. Cockerhams Male Labouring Tithables, Simon Gentry, Joseph Gentry, Richard Brooks, Elisha Brooks, Nicholas Gentry, Michael McKie Senr. & his Male Labouring Tithables, Thomas Low, Robert Wilson, William Wilson and Joshua Hawkins do forthwith clear and keep the same in Repair According to Law.

2 August 1757, Page 343
On the Petition of Clement Read Setting forth that he hath at his own private expense Built a Bridge over Little Roanoke River, and Open'd and Clear'd a Road from the Road that leads from the new Town over the County Bridge just above his Plantation to the Bridge Built by him and from thence to the Road down to the Court House and Church, which is since become a very Public Way, Praying that the Persons using the same that live convenient may be appointed to keep it in Repair, as also the Way from the Bridge above his House to the Office, and that they may be exemted from other Roads, thereupon (It Appearing that the said Road & Way is Convenient) It is Ordered that the said Clement Reads Male Labouring Tithables, James Taylors Male Labouring Tithables John Adkins's Male Labouring Tithables, Samuel Johnsons male Labouring Tithables, William Johnson, Valentine Austin and John Pleasants Male Labouring Tithables, do forthwith Clear & keep the said Road laid open by the said Clement Reads, and the Way from the Bridge above his House to his Office in Repair According to Law

3 August 1757, Page 348
Nehemiah Frank a Surveyor of a High Way in this County, having been Presented by the Grand Jury for not keeping in Repair the Road leading from the Horspen Creek to William Kays Plantation, having been duly Summon'd & not Appearing, It is Considered and Accordingly Ordered

that he make his fine with our Sovereign Lord the King by the Payment of fifteen Shillings Current Money, to his Majestys use, and that he pay the Cost of these Proceedings, and it is Ordered that he be taken &c^a.

3 August 1757, Page 349
Philip Jones a Surveyor of a High Way in this County, having been Presented by the Grand Jury for not keeping in Repair the Road whereof he is Surveyor having been duly Summon'd & not Appearing, It is Considered and Accordingly Ordered that he make his fine with our Sovereign Lord the King, by the Payment of fifteen Shillings Current Money to his Majestys use and that he pay the Cost of these proceedings, and it is Ordered that he be taken &c^a.

3 August 1757, Page 349
Thomas Vernon a Surveyor of a High Way in this County, having been duly Summon'd to shew Cause why he do not keep the Road whereof he is Surveyor in Repair and not Appearing, It is Considered and Accordingly Ordered that he make his fine with our Sovereign Lord the King, by the Payment of fifteen Shillings Current Money, and that he pay the Costs of these proceedings, and it is Ordered that he be taken &c^a.

3 August 1757, Page 349
Ordered that if Mathew Marable will mend the Bridge Over Blew Stone Creek near his House, that the Road that Passes over the said Bridge be Established a Public Road.

3 August 1757, Page 349
Stephen Evens is Appointed Surveyor of the Road leading from this Court House to Blew Stone Creek, and it is Ordered that he together with the hands that formerly Assisted him on the said Road, do forthwith Clear and keep the same in Repair According to Law

4 August 1757, Page 349
Clement Read is Appointed Surveyor of the Road by him Cleared from the Road that leads through the new Town over the County Bridge just above his Plantation to the Bridge by him built and from thence to the Road down to the Court House & Church.

6 September 1757, Page 379
Ordered that Joseph Williams, George Walton, Tscharner Degraffenreidt, and George Wells, or any three of them being first Sworn, do View and Examine the way from Kings Road across Meherin River above Willinghams Bridges to Hampton Wades, a Report to the next Court the Conveniency or inconveniency thereof.

4 October 1757, Page 382
Henry Blagrave Junr. is Appointed Surveyor of the Road leading from the North Maherrin to the Middle Maherrin whereof Henry Blagrave Senr. was late Surveyor, and it is ordered that he together with the Hands that Assisted the said Henry Blagrave Senr. thereon, do forthwith Clear and keep the same in Repair according to Law.

4 October 1757, Page 383
John Speed Gent^l is Appointed to Settle and divide the hands between Joshua Mabry and William Riddle, on the Roads whereof they are Surveyors

4 October 1757, Page 389
Ordered that William Hill, Thomas Laneir, Thomas Hawkins and John Speed or any three of them, direct the way for a Road to be Cleared from the Mine to the Road that goes to the Point &c^a

5 October 1757, Page 395
Mathew Marable is Appointed Surveyor of the Road leading from Blew Stone Creek to this Court House, whereof Stephen Evens was late Surveyor, and it is Ordered that he together with his own male Labouring Tithables, William Marables Male Labouring Tithables Branch Tanners Male Labouring Tithables and Joseph Hunt do forthwith Clear and keep the same in Repair according to Law.

6 October 1757, Page 401
Ordered that the Old Road that goes by Daniel Humphris's be Dropped and that the new Road that goes thro' the new Town be Established a Public, and it is Ordered that Thomas Bouldin and James Taylor Gent^l do Proportion the Hands to the Several Surveyors thereof from Little Roanoak Bridge to the old Road as they shall think fit.

6 October 1757, Page 408
Bryant Cooker is Appointed Surveyor of the Road to be Cleared the best and most Convenient Way from the Stony Hill by Cornelius Cargills to Kings Road, and it is ordered that he together with such hands as William Goode Gent[l] shall Appoint do forthwith lay open Clear and keep the same in Repair according to Law, and it is Ordered that the said Goode do Appoint such hands to assist him as he shall think fit.

6 October 1757, Page 408
Ordered that the Male Labouring Tithables at John Pleasants Quarter in this County and Henry Isbell and his Male Labouring Tithables do forthwith Assist Clement Read in Clearing and keeping the Road whereof he is Surveyor in Repair.

6 October 1757, Page 408
William Almon is Appointed Surveyor Randolphs Road and it is Ordered that he together with the hands that John Harvy shall Appoint do forthwith Clear and keep the same in Repair According to Law, and it is Ordered that the said John Harvy do Appoint such hands to Assist him as he shall think fit, and it further ordered that the said Harvy do Appoint such hands as he thinks fit to Assist Christopher Parsons in Clearing and Keeping in Repair the Road whereof he is Surveyor.

1 November 1757, Page 3
Henry Isbell is Appointed Surveyor of the Road whereof Own Sullivant was late surveyor, and it is Ordered that he together with the hands that Assisted the said Sullivant thereon, he Own hands and Joel Towns's hands do forthwith Clear and keep the same in Repair According to Law.

1 November 1757, Page 3
Ordered that a Road be laid Open and Cleared the best and most Convenient Way from the Road opposite to Abraham Martins House to Kings Road, and the said Abraham Martin is Appointed Surveyor thereof, and it is Ordered that he together with his Own Male Labouring Tithables, Abraham Vaughn, Robert Davis and his Son, Benjamin Adkins, Richard Ellis, William Ellis, John Sansum, Joshua Wharton, John Silcock, Edward Harris & William Gillum, do forthwith lay open Clear and keep the same in Repair According to Law.

1 November 1757, Page 3
Ordered that Robert Woods together with Osborn Keeling, Wilson Mattox, James Orr, Ensworth Midleton, Richard Berry, George Hannah, William Adams, James Adams, Nathan Adams, John Waller, Robert Martin, Thomas Price, Joel Stow, John Orr, Samuel David, James Caldwell, John Caldwell, George Scot, Perrin Alay, John Leeper, John Templin, David Kenedey, and James Murphy do forthwith Clear and keep the Road in Repair According to Law whereof the said Woods Surveyor.

1 November 1757, Page 3
Joshua Charin is Appointed Surveyor of the Road whereof Nehemiah Frank was late Surveyor, and it is Ordered that he together with the Hands that Assisted the said Frank thereon, do forthwith Clear and keep the same in Repair According to Law.

1 November 1757, Page 3
Richard Blanks is Appointed Surveyor of the Road whereof Henry May was late Surveyor, and it is Ordered that he together with the Hands that Assisted the said May thereon, do forthwith Clear and keep the same in Repair According to Law.

1 November 1757, Page 4
Thomas Hawkins, William Hill & Thomas Lanier three of the Persons Appointed by an Order of the last Court to View the Way for a Road from the Mine to Cocks Creek, this day Returned their Report thereon which is Ordered to be Recorded, and William Ballard is Appointed Surveyor thereof, and it is Ordered that he together with Stephen Hatchel, John Goin, William Goin, John Goin, Junr. William Glading, Stephen Mallet Junr. and John Ruffins Male Labouring Tithables do forthwith lay open, Clear and keep the same in Repair According to Law.

1 November 1757, Page 4
Ordered that a Road be laid Open and Cleared the best and Convenientest Way from Cocks Creek to Mizes's Ford, and Isaac Holmes is Appointed Surveyor thereof, and it is Ordered that he together with Henry Delonys Male Labouring Tithables, John Ballards Male Labouring Tithables John Watsons Male labouring Tithables, James McCoy and Henry Tally do forthwith lay Open Clear and keep the same in Repair According to Law.

1 November 1757, Page 4
Samuel Bugg is Appointed Surveyor of the Road leading from Allens Creek to Butchers Creek, whereof Thomas Easland was late Surveyor and it is Ordered that he together with the hands that Assisted the said Easland thereon do forthwith Clear and keep the same in Repair According to Law.

1 November 1757, Page 4
William Harris and John Wilborne, being two of the Persons Appointed by this Court to View the Way for a Road from the Woodpecker Creek to Col° Birds Mill, this day Returned a Report thereon, which is Ordered to be Recorded and John Cox is Appointed Surveyor thereof and it Ordered that he together with Edward Willis and his hands, John Flin, Thomas Flynn, John Tomson, Jacob Royster and his hands, Loflin Flins hands and John Roberts do forthwith lay Open Clear and keep the same in Repair According to Law.

1 November 1757, Page 4
Henry Gill is Appointed Surveyor of the Road whereof David Garland was late Surveyor, and it is Ordered that he together with the hands that Assisted the said Garland thereon do forthwith Clear and keep the same in Repair According to Law.

1 November 1757, Page 4
Thomas Watkins, Michael Prewet and Richard Prewet, being three of the Persons Appointed by this Court to View the Way for a Road from Hunts Ferry to the County line, this day Returned their Report thereon and Ordered to be Recorded. and it is Ordered that Charles Hunt with his hands do lay open the said Road until there shall be a Surveyor Appointed.

1 November 1757, Page 4
Zackariah Baker is Appointed Surveyor of the Road whereof John Glass was late Surveyor, and it is Ordered that he together with the hands that Assisted the said Glass thereon, do forthwith Clear and keep the same in Repair According to Law.

1 November 1757, Page 4
John Ragsdale is Appointed Surveyor of the Road whereof William Wallice was late Surveyor, and it is Ordered that he together with the hands that Assisted the said Wallice thereon, do forthwith Clear and keep the same in Repair According to Law.

1 November 1757, Page 5
Ordered that John Hix together with Mrs Cockerhams Male labouring Tithables, Simon Gentry, Joseph Gentry, Richard Brooks, Elisha Brooks, Nicholos Gentry, Michael M^ckie and his Male labouring Tithables, Thomas Low, Robert Wilson, William Wilson, Joshua Hawkins, William Embrys Male labouring Tithables, Robert Brooks, John Trice, Stephen Crump, and William Brooks's Male Labouring Tithables, do forthwith Clear and keep in Repair the Road whereof the said Hix is Surveyor, and it is Ordered that when the said Road is Clear'd that William Embry do Appoint such hands as he shall think fit to Work on the said Road with the said John Hix.

1 November 1757, Page 5
On the Petition of Thomas Hawkins and others for a Road to be Cleared the best and Convenientest Way from Butchers Creek at Akins's Ford to Allens Creek, into the Road that leads from Capt. Michells Ferry to the Point, it is Ordered that the said Thomas Hawkins, Pinkethman Hawkins and Thomas Moore being first Sworn, do View and Examine the Way where such Road is intended to be Cleared, and Report to the next Court the Conveniency or inconveniency thereof.

2 November 1757, Page 12
Ordered that Thomas Bouldin together with James Sullivant Thomas Portwood and Menoah Sullivant together with the hands formerly under him on the Road whereof he is Surveyor, do forthwith Clear and keep the said Road in Repair According to Law.

3 November 1757, Page 20
Ordered that Thomas Bouldin, James Taylor and Thomas Bedford Gent^l. direct the Turning the Road in such Places as shall appear to them to be Necessary, from the said Bouldins to Whites Plantation, and it is further Ordered that they Proportion and Divide the hands Convenient to the said Road and Jones' Road, Between the Surveyor of the said Road and Jones's Road, and it is further Ordered that the Surveyor of Jones's Road with Such hands as the said Thomas Bouldin, James Taylor and Thomas Bedford shall Appoint do forthwith Clear and keep in Repair the Road Leading from Comers Road to Prince Edward County Line

6 December 1757, Page 21
Ordered that the Road lately laid open and Cleared by David Caldwell from his House into Coles Road be kept in Repair by the said Caldwells own hands.

6 December 1757, Page 23
Thomas Foster is Appointed Surveyor of the Road whereof Philip Jones was late Surveyor, and it is Ordered that he together with the hands that Assisted the said Jones thereon, do forthwith Clear and keep the same in Repair According to Law.

6 December 1757, Page 23
Ordered that Henry Cox, Godfrey Jones, Daniel Hankins, and Stephen Collins, or any three of them. (being first Sworn &c^a) View and Examine the way for a Road to be Cleared .... to Turn our of the Road near the new Town, to Strike Godfrey Jones's Roling Road, and Report to the next Court the Conveniency or inconveniency thereof.

6 December 1757, Page 23
Ordered that William Embry, Daniel Claiborne, and William Goode Gent^l View and Examine the Road leading from the Mine into the Road to the Point, and Report to the next Court, the Conveniency or inconveniency thereof.

6 December 1757, Page 23
Thomas Covington is Appointed Surveyor of that part of the Road whereof Samuel Comer was late Surveyor leading from John Williams's Plantation downwards

6 December 1757, Page 23
Thomas Portwood is Appointed Surveyor of that part of the Road whereof Samuel Comer was late Surveyor leading from John Williams's Plantation upwards.

7 December 1757, Page 28
Richard Coleman is Appointed Surveyor of the Road whereof William Hawkins was late Surveyor, and it is Ordered that he together with the hands that Assisted the said Hawkins thereon, do forthwith Clear and keep the same in Repair According to Law.

8 December 1757, Page 30
County Levy
To James Hunt for keeping Cubb Creek Bridge in Repair five Years ... 9.0.0

7 March 1758, Page 35
William Hunt is appointed Surveyor of the Road whereof John Wilborne was late Surveyor. And it is Ordered that he together with the Assistance that Assisted the said Wilborne on the said Road do forthwith Clear & keep the same in Repair According to Law.

7 March 1758, Page 35
Godfrey Jones Henry Cox And Daniel Hankins Three of the Persons appointed by an Order of the Last Court to View and Examine a way for a Road to Lead from the New Town to Godfrey Jones's Roling Road. This day Returned a Report thereon, & thereupon the same is Ordered to be Recorded. And the said Daniel Hankins is appointed Surveyor Thereof. and it is Ordered that he together with George Foster Junior Thomas Taylor, Stephen Collins, and his Male Labouring Thiths, Thomas Nashes Male Labouring Tiths on Spring Creek, Clement Read Jun^{rs}. Male Labouring Tiths Godfrey Jones's male Labouring Thiths, John Russell John Hany John Foster, George Foster's Male Labouring Tiths, John Smith Daniel Hughbank, & Moses Hall do forthwith Clear & keep the same in Repair According to Law.

7 March 1758, Page 37
William Royster is appointed Surveyor of the Road whereof William Richardson was late Surveyor And it is ordered that the said Royster with the Assistance that assisted the said Richardson on the said Road do forthwith Clear and Keep the same in Repair According to Law.

7 March 1758, Page 38
On the Motion of George Wells leave is granted him to Clear a Road from Owls Creek to Moores Road.

7 March 1758, Page 38
Richard Witton Gent is appointed to appoint Such Hands as he shall think fit to Assist John Hawkins in Clearing and Keeping in repair the Road ordered to be Cleared in March Court one Thousand Seven hundred and fifty five Whereof the said John is Surveyor.

8 March 1758, Page 39
On the Petition of sundry the Ihabitants of Little Roanoak River for a Bridle Way to be Cleared the best and Convenientest way from the Mosing ford on the said Little Roanoak River to the Plantation of the Reverend M^r. William Kay Deceased It is ordered that David Caldwell, Joseph Perrin, And Thomas Ligan, being first Sworn do View and Examine

a way there such Bridle way is to be Cleared, and Report to the next Court the Conveniences or unconveniences thereof.

8 March 1758, Page 44
Ordered that the Road that goes over Willinghams Bridges be Droped & that all Surveyors heretofore appointed thereon be Discharged from working thereon.

8 March 1758, Page 44
Ordered that the Road that Crosses Little Roanoak Bridge above Clement Reads be dropt & that the Surveyors formerly appointed on the said Road be Discharged from Working on the same and that the Hands that assisted Thomas Bouldin Gent. late Surveyor of Part of the said Road be added to Samuel Comer's Hands on the Road whereof the said Samuel is Surveyor.

8 March 1758, Page 44
Ordered that Robert Woods with the Hands formly appointed to Assist him do Clear & keep in Repair the Road whereof he is Surveyor as far as the New Town.

8 March 1758, Page 44
Ordered that Clement Read with the Hands Formily appointed to Assist to Clear and keep in Repair the Road whereof he is Surveyor as far as the New Town & further that the Hands that Assisted Samuel Johnson in Clearing & keeping in Repair that Part of the Road whereof the said Samuel was late Surveyor that Crosses Little Roanoak above the said Clement Reads, be added to the Hands formerly appointed to Assist the said Read in Clearing and keeping in Repair the Road Whereof the said Reed is Surveyor.

8 March 1758, Page 44
Ordered that a Road be Laid open & Cleared the Best & Convenientest Way from the New Town to the Road that Crosses Little Roanoak near Joseph Mortons & that James Taylor Gent appoint such Hands as he Shall think fit to lay open and Clear the same.

8 March 1758, Page 45
On the Petition of Sundry Ihabitants of this County Ordered that a Bridle way be laid open & Cleared the best and Convenientest way from

Pinkethman Hawkins to Jeffersons Church. & William Ballard is appointed Surveyor of the said Road. And it is Ordered that he together with Robert Coleman, Christopher Coleman, Pinkethman Hawkins, Hugh Norwell, Thomas Norwell, John Goin, Henry Jackson, John Coleman, Thomas Farrar, James Coleman Thomas Draper, & Thomas Moore do forthwith lay open Clear & keep the same in Repair According to Law.

10 March 1758, Page 61
Joseph Williams Gentleman is appointed to appoint such hands as he shall Think fit to lay open Clear and keep in Repair a Road from Breedloves Creek to George Moores the way that is Marked by George Wills & John Foster if the said Williams shall think Convenient.

10 March 1758, Page 61
Ordered that Ruben Morgan turn the Road whereof Julius Nicholas is Surveyor on John Patricks Land the Most Convenientest way from Dorches Creek to John Taylor Dukes.

4 April 1758, Page 63
Ordered that Charles Hunt together with the following Persons to wit James Hunts hands, Matox Mayes & his William Black Josiah Holmes & his, William Mayes & Joseph Bayes, Richard Bookers hands, John Fuquay & his William Dickerson & James Cunningham do forth with Clear and keep the Road whereof the said Charles is Surveyor in Repair According to Law.

4 April 1758, Page 63
Ordered that Jacob Roberson together with the following persons, to wit, William Fuquay & his, Joseph Fuquay, Richard Sullins Thomas Boughton, John Boughton, John Kiersey Thomas Keirsey, John Kiersey j^r. William Eastland & his, John Butler, James McGlaughlin, Robert Andrews & his, Agnus Campbell William Hardwick, William Cunningham & his, Peter Young, Joseph Davis Thomas Roberts, Peter Franklin, Samuel Lafon, Andrew Rodgers John Rodgers, Alexander Roberson, Thomas Moore & his, David Logan & his, and Leonard Keelin do forthwith Clear & keep the Road whereof the said Jacob is Surveyor in Repair According to Law.

4 April 1758, Page 64
Ordered that Thomas Rodgers together with the following Persons, to wit, Alexander McConnell & his, Thomas Daugherty & his, James

Daugherty James Burnsides, James Mitchell John Smith & his, John McNess, John Greer, James Franklin William Thompson, William Standfield & Thomas Standfield do forthwith Clear, & keep the Road where of the said Thomas is Surveyor in Repair According to Law.

4 April 1758, Page 64
Ordered that David Caldwell together with the following Persons, to wit, James Little William Dudgeon & his John Vance Jacob Jones, John Dudgeons, John Wood, William Caldwell John East, Thomas East, Francis Mann, Page Mann, James Barton & his, John Clark, Richard Adams, Arthur Slaton, & Daniel Slaton, James McMurday, James Cunningham & his, Robert Caldwell & his William Philby, William Hall John Caldwell, John Fulton & Alexander Berryhill do forthwith Clear and keep the Road whereof the said David is Surveyor in Repair According to Law.

4 April 1758, Page 64
Ordered that Richard Dudgeon together with the following Persons to wit Seth Ward, & his, John Ward & his, William Dudgeon & M^{rs}. Gladses Hands, Tarrance McDaniel, Isaac Vernon, & his, Thomas Kersey Senior, James Vernon, & Jonathan Vernon, do forthwith Clear and keep the Road whereof the said Richard is Surveyor in Repair According to Law.

4 April 1758, Page 64
John Stewart is appointed Survey of the Road in the Room of Thomas Stewart It is ordered that he together with the Hands that worked on the said Road under the said Thomas Stewart, do forth with Clear & keep the same in Repair According to Law.

4 April 1758, Page 65
Ordered that John George Penington together with John Mize John Bates, John Ezill, Michael Ezill, William Pennil, Robert Conel, & George Chaves do forth with Clear and keep the Road in Repair According to Law whereof the said Pennington is Surveyor.

4 April 1758, Page 68
John Ezell is appointed Surveyor of the Road where of Joshua Mabry was late Surveyor, leading from the Piney Ponds to Inghrams Road, And Henry Delony and John Speed Gent. are appointed to appoint such Hands as they shall think fit to assist the said John Ezell in Clearing and keeping the said Road in Repair According to Law.

4 April 1758, Page 68
John Speed and Henry Delony Gent^l. are appointed and Desired to appoint such hands as they shall think fitt to assist William Riddle in Clearing & keeping the said Road in Repair according to Law, whereof the said William is Surveyor.

4 April 1758, Page 68
John Foster is appointed Surveyor of the Road leading from George Moores to the great Owl Creek, And it is ordered that he together with the following male Labouring Tithables, to wit, John Nance, John Cole, Richard Nance, William Nance, Elijah Wells, Fredrick Nance, John Foster, George Wells, & George Walton & his, do assist the said John in Clearing and keeping the said Road in Lawfull Repair

4 April 1758, Page 69
Barnabus Wells is appointed Survey of the Road leading from the great owl Creek the Best and Convenientest way to Breedloves fork of the Juniper, and it is ordered that he together with the Male Labouring Tithables, to wit, James Roberts Senior, James Roberts J^r. Barney Wills, Thomas Thornton, Thomas Shelborne, Benjamin Wilks, Argil Baxton, Willian Irvin, Peter Petty Poole, Thomas Crenshaw, Cornelius Crenshaw, James Haily, John Haily, Edward Haily And John Wells do assist him in Clearing and keeping the Road in Repair According to Law.

2 May 1758, Page 76
Ordered that John Hawkins together with the following Persons, to wit Jeremiah Hatcher & his, Hands, Nicholas Hobson & his, John Hobson & his, William Wilson, Robert Wilson, William M^c. Doe(?), Michell Hawkins, Joseph Boswell & his, Edward Goode & his, George Elliott & his, Thomas Pettie & his, Philip Poindexter & his, & John Cockerham & his do forwith clear & keep the road in repair, whereof the s^d. Hawkins is Surveyor according to Law.

2 May 1758, Page 76
Ordered that George Elliott, together with the following Persons, to wit, William Stone & his Hands, John Chandler, William Lax, Obediah Hooper do forthwith clear & keep in repair the road leading from Twithys old road, to John Hawkins's road, whereof s^d. Elliott is Surveyor according to Law.

2 May 1758, Page 77
Grand Jury Presentments
...the Surveyor of the road from Mossin Ford to Capt. Thomas Bouldin's in Cornwall Parish
...the Surveyor of the new Piece of road round Mr. Marrable's Plantation, Part in Cumberland Parish & Part in Cornwall Parish ...

6 June 1758, Page 91
Mathew Marable a Surveyor of a Highway in this County Who stands presented by the Grand Jury, for not keeping the same in Repair appeared and on hearing the arguments on both sides, it is ordered that the said Presentment be dismissed.

6 June 1758, Page 91
Henry Isbel a Surveyor of a Highway in this County, who stands presented for the Grand Jury for not keeping the same in Repair appeared, and on hearing the Arguments on both sides, it is considered that the said Presentment be dismiss'd.

19 July 1758, Page 104
A Petition of sundry Inhabitants of this County, praying a Ferry may be kept Over Roanoake River at Francis Wagstaff's Landing in this County is presented, read and held to be reasonable, which is ordered to be certified to the general Assembly.

1 August 1758, Page 108
Robert Wilson is appointed Surveyor of the road whereof Henry Cockerham was late Surveyor And it is ordered, that he together with the Hands that assisted the said Cockerham thereon, do forthwith clear and keep the same in repair according to Law.

1 August 1758, Page 109
On the Motion of Daniel Claiborne Gentleman Leave is granted him to cut a Bridle Way from Willingam's road, the nearest & best Way to Randolph's Road.

5 September 1758, Page 111
Daniel Malone is appointed Surveyor of the Road leading from Willingham's Road to Roberts Waggon road, and Lyddal Bacon Gentleman

is desired and appointed to proportion such Hands as he shall think fit to assist the said Daniel in clearing and keeping the said road in repair according to Law.

5 September 1758, Page 111
Lyddal Bacon Gentleman is appointed and desired to appoint such Hands as he shall think fit to assist William Petty Pool in clearing and keeping the road in repair, whereof the said William is Surveyor.

5 September 1758, Page 113
Ordered that William Caldwell & James Hunt Gent treat with Workmen to build a Bridge over Turnip Creek in what Manner, and for what Consideration they shall think proper.

5 September 1758, Page 114
Ordered that Thomas Bouldin and Thomas Bedford Gentlemen treat with Workmen to build a Bridge over little Roanoak river above Clement Read's in what Manner and for what Consideration they shall think proper.

3 October 1758, Page 115
On the Petition of David Caldwell for Leave to turn the road near George Pattitors it is ordered that John Waller Richard Dudgeon & John Logan (they being first sworn before a Justice of this County) do view the best thereof, and make a report of their View here to the next Court.

7 November 1758, Page 116
David Gwyn is appointed Surveyor of the road whereof Henry Isbell was late Surveyor, and it is ordered that he together with such Hands as worked on the said Road with Isbell do forthwith clear and keep the same in repair according to Law.

8 November 1758, Page 121
County Levy
To Thomas Bouldin & Thomas Bedford for a Bridge over little roanoak river near Col: Clement Read's   14.15.0
To James Mitchel for a Bridge over Turnip Creek ...   8.0.0

To David Caldwell for a Bridge over Turnip Creek   1.5.0

3 April 1759, Page 132
Thomas Bouldin Gentleman is appointed Surveyor of the road from little Roanoke Bridge to Randolph's Road and it is ordered that the usual Hands assist him in clearing and keeping the same in repair according to Law.

3 April 1759, Page 133
On the Petition of David Bridgforth for a road to be clear'd the best and most convenient Way through Hugh Lawson's Land in his old rooling way as he formerly made Use of. It is ordered that Daniel Wynne Joel Ferguson, and Daniel Mays or any two of them they being first sworn before a Magistrate of this County do view and examine the Way whereof such road is intended to be cleared and report to the next Court the Conveniences and inconveniences.

3 April 1759, Page 133
On the Petition of Sundry Inhabitants of Cubb Creek in this County to turn the road that run through the Plantation of William Caldwell Gentleman It is ordered, that William Dudgeon John East and Francis Mann or any two of they they being first sworn before a Justice of this County do view and examine the Way thereof the said road is intended to be built and report here to the next Court the Conveniences and inconveniences thereof.

3 April 1759, Page 133
Ordered that the several Surveyors of the road in this County be continu'd 'till the next Court to be held for this County.

1 May 1759, Page 138
The Persons appointed by this Court, to view the Way petition'd to turn the road round William Caldwell's Plantation this Day returned their report and it is order'd that the said road be established according to the said petition, and that the usual Hands which formerly worked on the said road assist in turning the said road thro' plantation of the said William Caldwell.

1 May 1759, Page 139
Ordered, that the Overseer of the road together with the Hands which usually assist him, forwith put in repair for public Use the road leading through the Plantation of Mathew Marrable Gent¹ according to Law.

1 May 1759, Page 139
Grand Jury Presentments
...the Surveyor of the road from the Court house to Mathew Marable's Gentlemen ...

3 May 1759, Page 159
Ordered, that a Bridge be built over the Midle Meherrin river at the Place where the Bridge was formerly built, and that Richard Witton & Joseph Williams Gentlemen treat with Workmen to build the same in such Manner and upon such Conditions as they think proper, and that the Undertakers give Bond and Security to the said Richard Witton & Joseph Williams to warrant and maintain the same seven Years.

3 May 1759, Page 159
It appearing to the Court, that Silvanus Walker hath not kept the Bridge in repair which he undertook to do according to the Condition of his Bond entered into for that Purpose, It is therefore ordered, that a Suit be entered against him to appear here at the next Court, to shew Cause why he did not keep the said Bridge in repair according to the Condition of the foresaid Bond.

3 May 1759, Page 159
It appearing to the Court, that Nenucan Hunt hath not kept the Bridge in repair which he undertook to do, according to the Condition of his Bond enter'd into for that Purpose, It is therefore ordered, that a Suit be entered against him, at appear here at the next Court, to shew Cause, why he did not keep the said Bridge in repair, according to the Condition of the aforesaid Bond.

6 June 1759, Page 6
John East, Francis Man, and William Dudgeon being three of the persons named by this Court to view the road from William Caldwell's Plantation, this day returned their report thereon which is ordered to be recorded.

6 June 1759, Page 9
Henry Williams is appointed Surveyor of the Road leading from this Court house to Blue Stone, the new Way clear'd by Mathew Marrable Gentleman and it is ordered, that the said Williams together the Male labouring Tithables convenient thereto do forthwith clear and keep the same in repair according to Law.

6 June 1759, Page 9
Isaac Johnson is appointed Surveyor of the road leading from Blue Stone to King's road, whereof Henry Williams and Jn°. Good were of late Surveyors, and it is ordered, that he together with Hands which usually assisted the said Williams & Goode on the said road do forthwith clear and keep the same in repair according to Law.

6 June 1759, Page 9
William Roberson is appointed Surveyor of the road leading from this Court House to Great Blue Stone, and it is ordered, that he together with the Male labouring Tithables convenient thereto do forthwith clear and keep the same in repair according to Law.

6 June 1759, Page 9
William Roberts is appointed by this Court a Ferryman at Blank's Ferry in this County, and it is ordered, that he be exempted from the payment of public and County Levies during his Continuance in such Employment.

3 July 1759, Page 15
On the Petition George Elliott to open and keep in repair the old road leading from this Courthouse crossing Meherrin at the said Elliot's House continuing the same to Reedy Creek Church, It is ordered, that Lyddal Bacon & Thomas Tabb Gent: being first sworn, do diligently view & examine the road petitioned for, and return a report thereof to the next Court.

3 July 1759, Page 15
Ordered, that a Bridge be erected and built over Allen's Creek, and John Speed and Henry Delony are appointed & desired to let the same to undertakers upon such Terms, & to be built in such Manner & Form as they in their Discretion shall think proper.

3 July 1759, Page 15
Ordered, that John Johnson, Isaac Mitchell and Nathaniel Robertson, or any two of them, being first sworn &c^a. do view & examine the road leading from King's Foard to the Country Line, and report the same here to the next Court.

3 July 1759, Page 15
Thomas Tabb Gentleman is appointed and desired by this Court to act in Conjunction with Mathew Marable Gentleman in the room of William Embry Gent dec^d. to join Abraham Cocke & Thomas Bowrey Gent Members of the Court of Amelia County to contract & agree with such Workmen as they shall judge proper to build a Bridge over Great Nottoway river from this County to the said County of Amelia as the Law directs

3 July 1759, Page 15
Richard Ship is appointed Surveyor of the road leading from Little roanoke Bridge to the Head of Bryery river, and it is ordered, that he together with the following assistance to wit, Josiah Moreton's Thomas Spencer's, Thomas Nash's and Richard Womack's Male labouring Tithables do forthwith clear and keep the same in repair according to Law.

7 August 1759, Page 18
On the Petition of Richard Blanks Licence is granted him to keep a Ferry on the river roanoke at Berry's Ford he giving Security; whereupon he together with the Mathew Marable Gentleman his Security entered and acknowledged their Bond according to Law for that Purpose.

7 August 1759, Page 18
The road petitioned for at the last Court by George Elliot is not to be continued as a publick road, but Leave is granted him to open and keep it in repair at his own Expence, if he thinks proper.

7 August 1759, Page 18
Ellick Joyce, Joseph Moreton Junr. and Clement Read Gentlemen are appointed and desired to view the road leading from the new Town to the upper little roanoke Bridge and report the Condition thereof to the next Court.

7 August 1759, Page 18
Joseph Williams and Thomas Tabb Gentlemen are appointed and desired to treat with Workmen about building two Bridges over the river Meherrin in such Manner and upon such Terms as they shall judge proper.

7 August 1759, Page 19
Jacob Royster and William Royster are appointed and desired to view a Way leading from the said William Royster's Ferry to the Country Line and report the Condition thereof here to the next Court.

8 August 1759, Page 21
Mathew Marable and Lyddal Bacon Gentlemen are appointed and desired by the Court to treat with Workmen about building a Bridge over the North Meherrin river at Silvanus Walker's Ford in such Manner and upon such Terms as they shall judge proper.

4 September 1759, Page 24
Edmund Taylor is appointed Surveyor of the Roads, whereof Nathaniel Robertson and Zachariah Baker were late Surveyors; and it is ordered that he together with the Assistance that assisted the said Roberson & Baker do forthwith clear and keep the same in repair according to Law.

4 Septemer 1759, Page 24
Chesley Daniel is appointed Surveyor of the road lately cleared by him and it is ordered, that he together with this own, Leonard Daniel's and John Johnson's Male Laboring Tithables do keep the same in repair according to Law.

4 September 1759, Page 25
Joseph Williams Gentleman is appointed and desired by the Court, to proportion the Hands, that work upon the Road leading from Roanoke river to the Fork of Randle's Road and appoint Overseers thereof.

5 September 1759, Page 27
Henry Delony and John Speed Gentlemen are appointed and desired by this Court, to treat with Workmen about building of a Bridge over Miles Creek at the usual Place in such Manner, & on such Terms as they shall judge proper.

5 September 1759, Page 30
Thomas Tabb and Lyddal Bacon Gentlemen are appointed to wait on the Court of Amelia County to inform the Court that they are appointed to contract and agree with Workmen for building a Bridge over great Nottoway river from this County to the said County of Amelia, and to require the said Court of Amelia to join with the said Tabb & Bacon in such Agreement as the Law directs, or to appoint proper persons so to do.

2 October 1759, Page 31
Alexander Joyce and Joseph Morton, two of the Persons appointed by this Court, to view the road leading from the New Town to the upper little Roanoke Bridge, this day returned a report thereon, which is ordered to be recorded.

2 October 1759, Page 32
Richard Palmer is appointed Surveyor of the road, whereof Thomas Satterwhite was late Surveyor, and it is ordered, that he together with the Hands that assisted the said Satterwhite forthwith clear and keep the same in repair according to Law.

2 October 1759, Page 33
John Green is ordered (together with the Male labouring Tithables) to clear a road from Aaron's Creek below Peter Overbey's to his Ferry, and it is further Ordered, that he keep the same in repair according to Law.

2 October 1759, Page 34
William Harris is appointed Overseer of the Road whereof William Hunt was late Overseer, and it is ordered, that he together with the Hands, that assisted the said Hunt clear and keep the same in repair according to Law.

6 November 1759, Page 38
Grand Jury Presentments
...the Surveyor for not keeping the Road in good Repair from Marables to the Court House ...
...the Surveyor of the Road from Palmers Ferry to Samuel Phelps's for not keeping the same in Repair ...

...the Surveyor of the Road where Kings road goes out of Randolphs Road for not Setting up a Post of Directions ...

...the overseer of the road from the said forks of Kings Road to the Parsons for not keeping the same in Repair ...

...the Surveyor of the Road from the Magazine to the Bridge on Little roanoke above Col°. Reads ...

...the Surveyor of the Road at the next forke above Richard Colemans for not Setting up a posts of Directions at the said Fork ...

...the Surveyor of the road at the fork between Richard Colemans and Reedy Creek Church for not Setting up a Post of Directions...

...the Surveyor of the Road from Flatt Rock Creek to great Creek Bridge ...

...the Surveyor of the Road from Meherrin River to washburnes Path at the forke of Coles Road ...

* * *

we also Present by the oath of Joseph Minor the Surveyor of the Road from the Head of Mountain Creek to notway bridge at ralphs Sheltons, a presentment against Thomas Lowry for driving of Horses on the sabath day ...

...the overseer of the Road from Little Roanoake Church to the Magazine ...

* * *

...the Surveyor of the Road from the Parsons to Marables ...

6 November 1759, Page 41

John Stewart is appointed Surveyor of the Road Leading from Stewarts Ferry to where Thomas Comer lives, And Paul Carrington & Henry Childress are appointed and Desired to nominate assistance to work under the said Stewart. And it is ordered that he together with Such assistance as they shall appoint, do forthwith Clear & keep the same in Repair According to Law.

6 November 1759, Page 41
Reese Hughes is appointed Surveyor of the Road leading from where Thomas Comer lives to the Mosing Foard Bridge, Paul Carrington and Henry Childress are appointed and Desired to Nominate Assistance to work under the said Hughes, and it is ordered that he together with such assistance as they shall appoint, do forthwith Clear & keep the same in Repair according Law.

6 November 1759, Page 41
Joel Townes is appointed Surveyor of the Road leading from Bouldins road Crossing Little Ronoak at Gwins Bridge and from thence to Marables Mill. Thomas Boulding and Thomas Bedford Gent. are Desired and appointed to Nominate the assistance to work under the said Towns, And it is ordered that he together with Such assistance as they shall appoint do forthwith Clear & keep the same in Repair according to Law.

6 November 1759, Page 42
William Adams is appointed Surveyor of the Road leading from Marables Mill to falling River Road Thomas Bouldin & Thomas Bedford Gent. are appointed and desired to Nominate the assistance to work under the said Adams. And it is ordered that he together with such assistance as they shall appoint, do forthwith Clear and Keep the same in Repair according to Law.

3 December 1759, Page 44
County Levy
To Samuel Bugg for Building a Bridge over Allens Creek ...
        22.19.0
To Joseph Williams for d°. over Middle Meherrin ...
        40.0.0

5 February 1760, Page 53
Lazarus Willims is appointed Surveyor of the Road from Ready Creek Church to Daniel Hayse, And it is ordered that he together with the male Labouring Tithable Persons, that Assisted the Late Surveyor on the said Road, do forthwith Clear & keep the same in Repair, According to Law.

5 February 1760, Page 53
Thomas Baughon is appointed Survey of the Road whereof Jacob Robinson was late Surveyor, And it is Ordered that he together with the male Labouring Tithable Persons that assisted the said Jacob on the said Road, do forthwith Clear & keep the same in Repair Accord^g. to Law.

5 February 1760, Page 54
Thomas Nance is appointed Surveyor of the Road Leading from Randolphs fork up to the Head of Reeses fork where Edward Harris's Path Crosses Kings Road, And it is ordered that he together with the following male Labouring Tithable Persons, towit, Hills and his, Petis's & his, and Richard Hix Thomas Foster, & Micajah Smithson, do forthwith Clear and keep the same in Repair According to Law.

5 February 1760, Page 54
William Thomason is appointed Surveyor of that part of the Road that leads from Reases fork up to the Plantation of the Reverend William Kay, deceased, And it is ordered that he together with the following Assistance, towit, Joseph Hunt junior Micajah Francis, Benjamin Smith, Richard and William Ellis, Thomas Smith, Edmond Ellis, & Thomas Hamlin, do forthwith Clear and keep the same in Repair According to Law.

5 February 1760, Page 54
Charles Allen is appointed Surveyor of the Road whereof Richard Blanks was late Surveyor, And it is ordered, that he together with the Assistance that Assisted the said Blanks on the said Road do forth with Clear & keep the same in Repair according to Law.

5 February 1760, Page 54
Amos Hix is appointed Surveyor of the Road whereof Joel Chandler was late Surveyor, And it is ordered that he together with the Male Labouring Tithable Persons that Assisted the said Chandler on the said Road do forth with, Clear & keep the same in Repair According to Law.

7 February 1760, Page 67
Bennet Holloway is appointed Surveyor of the Road whereof George Holloway was late Surveyor, And it is ordered that he together with the assistance that assisted the said George on the said Road, do forth with clear & keep the same in Repair According to Law.

7 February 1760, Page 67
Daniel Hayse is appointed Surveyor of the Road leading from the Said Hayse's to great Notway on Cox's Road, And it is ordered that he together with the usual assistance that worked on the said road do forthwith assist the said Hayse in Clearing & keeping the same in Repair According to Law.

7 February 1760, Page 67
John Fulton is appointed Surveyor of the Road whereof Joseph Morteon was late Surveyor, And it is ordered that he together with the assistance that Assisted the said Morton on the said Road, do forthwith Clear & keep the same in Repair According to Law.

7 February 1760, Page 67
On the Petition of Mathew Marable Gent. leave is granted him to Clear a Road from his House into Cox's Road below the South Meherrin

7 February 1760, Page 68
On the Petition of Joseph Williams Gent. leave is granted him to Clear a Road from the Wolf Pitt on Kings road to his House and from thence to Kings Road on Reeses fork.

6 March 1760, Page 79
William Mitchell is appointed Surveyor of the Road Whereof William Hill was late Surveyor, And it is ordered that he together with the assistance that assisted the said Hill on the said Road do forth with clear & keep the same in Repair According to Law.

6 March 1760, Page 80
Thomas Lanier is appointed Surveyor of the Road where of James Parrott was late Surveyor, And it is ordered that he together with the assistance that assisted the said James on the said Road, do forthwith Clear & keep the same in Repair According to Law.

6 March 1760, Page 86
Samuel Johnston is appointed Surveyor the Road Leading from the Magazine to Little Roanoak Brige above Clement Reads Plantation, and Ja^s Taylor Gent. is appointed & desired to Nominate assistance to work under the said Samuel, And it is ordered that he together with such assistance as the said Taylor shall appoint, do forth with Clear & keep the same in Repair According to Law.

6 May 1760, Page 89
Francis Clarke is appointed Surveyor of the road Where of Edmund Taylor was late Surveyor, And it is ordered that he together with that assisted the said Taylor on the said Road, do forthwith Clear and Keep the same in Repair According to Law.

6 May 1760, Page 89
Ordered that a Bridge be Erected & built over Louse Creek And William Caldwell and James Hunt are appointed and Desired to Lett the same to undertakers for Seven Years upon such terms, & to be built in such Manner and form, as they in their Discretion shall think fit & Convenient.

6 May 1760, Page 89
Ordered that a bridge be Erected & built over Wards fork and Thomas Bedford and Thomas Bouldin Gent[l]. are appointed & desired to Lett the same to undertakers upon for Several Years upon Such terms, and to be built in such manner and form as they in their Discretion shall think fitt & Convenient.

6 May 1760, Page 89
John Bracy is appointed Surveyor of the road whereof Francis Wagstaff was late Surveyor, And it is ordered that he together with the assistance that assisted the said Wagstaff on the said Road do forthwith Clear and keep the same in Repair According to Law.

6 May 1760, Page 94
Grand Jury Presentments
...the Surveyor of the Road from the Robinson to Cocks road Called Waltons road ...
...the Surveyor of the Road from the Mossing ford Bridge to Bouldings ...
...the Surveyor of the Road from Blanks ferry to Thomas Fosters ...
* * *
...the Surveyor of the Road from Bouldings Church to the Magazine ...
...the Surveyor of the Road from Mountain Creek to Parrishes fford ...

6 May 1760, Page 96
Valentine Mullins is appointed Surveyor of the Road from the mouth of Butchers Creek to the Court House and it is ordered that he together with the following assistance, to wit, William Watkins, Mathew Avery, James Avery John Avery, Henry Avery, Mathew Carter, Luke Frolia, John Hayse, M^r. Armsteads Hands & Hezekiah Taber, do forthwith Clear & keep the same in Repair According to Law.

6 May 1760, Page 96
Richard Palmer is appointed Surveyor of the road from Palmers ford to Samuel Phelps, and it is ordered that he together with the following assistance towit, Robert Munfords Hands David Bullock, Robert Hood, James Parrott, James Wilkins, Edward Lewis, Robert Hester, Francis Bracy, George Bruce, Richard Long, William Donathan, & Richard Wilkins do forth with Clear & keep the same in Repair according to Law.

8 May 1760, Page 118
Order'd that a bridge be Erected & built over Cubb Creek at or near the Place called Dudgeon's ford, and William Caldwell & James Hunt Gent are appointed and desired to Lett the same to Undertakers upon such Terms, & to be built in such manner and form, as they in their Discretion shall think fit & convenient

8 May 1760, Page 118
Order'd that Henry Venduke, James Dause, and Samuel Rudder, (or any three of them being first sworn &c) do View & Examine a Way out of the Road Leading from Francis Rays to John Jennings & from thence to flat Rock Church, and Report to the next Court the Conveniency or inconveniency thereof

8 May 1760, Page 118
Order'd that a bridge be Erected & built over blue Stone at a place called Cargills Road, and Cornelius Cargill, and William Goode Gent are appointed and desired to lett the same to undertakers for Seven Years, & to be built in such manner & form as they in their Discretion shall think fitt & Convenient

8 May 1760, Page 118
Richard Witton and Thomas Tabb Gent^l. are appointed and desired to lett the Building or Repairing of the Bridge over the north Meherrin, called Hatchers Bridge, to undertakers upon such Terms as they in their Discretion shall think fit & Convenient

3 June 1760, Page 126
George Estis is appointed Surveyor of the Road, whereof Charles Allen was late Surveyor, and it is ordered that he together with the assistance that assisted the said Charles Allen on the said Road, do forth with Clear & keep the same in Repair according to Law.

3 June 1760, Page 127
Elisha Brook is appointed Surveyor of the Road Leading from John Humphris's to Joseph Bozwells, & It is ordered that he together with the assistance Convenient thereto do forth with Clear and keep the same in Repair According to Law.

3 June 1760, Page 128
David Gwin who Stands Presented by the grand jury for not keeping the Road in Repair whereof he is Surveyor, hearing the arguments on both sides, It is ordered that the said Presentment be Dismised

3 June 1760, Page 128
William Robinson is appointed Surveyor of the road lead^g. from the Court House to Wades Ferry, And it is ordered that he together with the following assistance, towit, The male labouring Tithable Persons belonging to Cornelius Cargill Gentleman John Cargill and his, John Camp and his, William Harris (Finney wood) and his, and the said Robersons hands, do forthwith Clear & keep the same in Repair According to Law.

1 July 1760, Page 146
James Sullivant is appointed Surveyor of Willinghams Road leading from the fork of the said Road, to the Bridges over Meherrin & Swish Creek, and Lyddal Bacon and Daniel Claiborne Gent^l are appointed and desired to Nominate assistance to work under the said Sullivant, and it is ordered that he together with such assistance as they shall appoint, do forth with Clear & keep the same in Repair According to Law.

1 July 1760, Page 146
Jonathan Davis is appointed Surveyor of Willinghams Road Leading from Swish Creek bridge to Roberts's waggon road And Lyddal Bacon, and Daniel Claiborne Gentl are appointed and desired to Nominate assistance to work under the said Davis and it is ordered that he together with such assistance as they shall appoint, do forth with Clear & keep the same in Repair according to Law.

1 July 1760, Page 150
James Taylor and Joseph Morteon Gent¹ are appointed & Desired to Lett the building or Repairing of the Bridges over Little Roanoak (called Morteons Bridge,) to undertakers upon such Terms as they in their Discretion shall think fit & convenient

1 July 1760, Page 150
Richard Hix is appointed Surveyor of the Road whereof Thomas Covington was late Surveyor, and it is ordered that he together with the assistance that assisted the said Covington on the said Road, do forthwith Clear and keep the same in Repair according to Law.

5 August 1760, Page 152
Ordered that the present Collectors for this County do forthwith Pay Edmund Ballard the sum of Six Pounds ninteen shillings & Three Pence half Penny Current Money for the Building of a Bridge over Nottoway River

5 August 1760, Page 153
On the Petition of John Cargill Praying leave to turn the Road from blue Stone Bridge into Kings road, Cornelius Cargill and John Cargill are appointed & desired to View and Examine the way where Such road is intended to be turned, and Report to the next Court the Conveniency or in Conveniency there of

5 August 1760, Page 154
On the Petition of Richard Rockett leave is granted him to keep a Bridle way along the old Road from the County Line to Langlys Store

5 August 1760, Page 154
On the Petition of Thomas Pettis and others Leave is Granted them to make a Bridle way through the Land of George Elliotts and John Robinson

5 August 1760, Page 155
Ordered that the Present Collector for this County do forthwith Pay William Chandler the sum of fifty Pounds Current Money for building a Bridge over the Meherrin above Scotts.

5 August 1760, Page 155
On the Petition of John Parker for a Road to be laid off & Cleared the best & most Convenientest way from the Round Meadow on Stoney Creek into flat rock road, It is ordered that Mathew Williams, William Johnson, and William Bean (or any two of them being first Sworn &c) do View & Examine the way where such road Petitioned for & report to the next Court the Conveniency or Inconveniency thereof.

5 August 1760, Page 159
Ordered that a bridge be Erected & built over the north Meherrin at Hawkins's, and Richard Witton and thomas Tabb Gent^l are appointed and Desired to Lett the same to undertakers upon such Terms, and to be built in Such Manner and form, as they in their Discretion shall think fit and Convenient.

6 August 1760, Page 169
David Gwin is Continued Surveyor of that part of the Road whereof he was late Surveyor, leading from the Mossing ford Bridge to Bouldins Path, and it is ordered that he together with the usual assistance that worked on the said Road do forthwith Clear and keep the same in Repair according to Law

2 September 1760, Page 170
On the Petition of Sundry of the Inhabitants of this County for a road to be Laid off and Cleared the best and mosts Convenient way from flat Rock Creek to Brunswick line by Mason Bishops And it is ordered that John Callingham be overseer of the said Road, and it is further ordered that he together with the following assistance, to wit, Samuel Rudd, David Callinghame, James Denton Thomas Hill, Arthanatious Elmore, William Pearson, Richard Wyatt, Robert Garrett, John Green, Nicholas Callingham, Humphrey Garrett, James Garrett, William Barker, William Baily, Thomas Trammell, John Parish, Joseph Taylor, Epharim Atkinson, Burrell Atkinson, Mathew Lafoone, and James Dukes, do forthwith lay open Clear and keep the same in Repair according to Law.

2 September 1760, Page 172
On the Petition of Sundry of the Inhabitants of this County for a Road to be Laid off and Cleared the best and most convenient Way from Captain Hunts Road into the road at Gryme's foard, and from thence into Buffelo Road, It is ordered that Robert Weakley, Richard Adams, James Franklin and Arthur Slayton, (being first Sworn &c) do diligently View and Examine the way where Such Road is Petitioned for, and Report to the next court the Conveniency or inconveniency thereof.

3 September 1760, Page 184
Gabriel Toombs is appointed Surveyor of the Road whereof Thomas Foster was late Surveyor, and it is ordered that he together with the assistance that assisted the said Foster on the said Road, do forthwith Clear and keep the same in Repair according to Law.

3 September 1760, Page 184
Order'd that a Bridge be Erected and built over blue Stone Creek and Cornelius Cargill Gentleman is appointed and desired to Lett the same to undertakers upon such terms and to be built in such manner and form, as he shall think fit and Convenient.

18 September 1760, Page 185
A Petition of Sundry Inhabitants of this County Praying a Ferry may be keept over Roanoake River from the Land of William Royster to the Land of William Byrd Esqr. is Presented and Read, and held to be Reasonable, which is ordered to be Certifyed to the General Assembly

4 November 1760, Page 187
On the Petition of Sundry of the Inhabitants of this County for a Road to be laid off and Cleared the best and most Convenient Way from Stokes Bridge to flat rock Church, It is ordered that Charles Parish, Reps Jones, Thomas Chambers, and Robert Dyer, or any three of them being first Sworn &c, do diligently View and Examine the Way where Such Road is Petitioned for, and Report to the next Court the Conveniency

4 November 1760, Page 187
On the Petition of Sundry of the Inhabitants of this County for a road to be laid off and Cleared the best and most Convenientest Way from Grasty Creek to Roysters ferry, It is ordered that Samuel Wilson, Edward Colbreath, and Peter Colbreath, being first Sworn &c do diligently Examine and View the way where such Road is Petitioned for, and Report to the next Court the Conveniency &c

4 November 1760, Page 188
On the Petition of John Howell and others for a Road to be laid off and Cleared the best and most Convenient way from Meherrin River near the mouth of flat Rock, from thence Crossing Stoney Creek and leading

into flat rock Road, It is ordered that William Beal, Matthew Organ and John Hight being first Sworn before a Majistrate &c, do diligently View and Examine the way where Such Road is Petitioned for, and Report to the next Court the Conveniency &c

4 November 1760, Page 189
Grand Jury Presentments
...the Keeper of the upper Bridge on Cub Creek ...
...the keeper of the Bridge called Cox's ...
...the overseer of the Road from the Hither Juniper to Cox's Road ...
...the Overseer of the Road from the Parsons Barn to the Court House ...
...the Survey of Kings road from the Parsons Barn to Randolphs Road ...
...the Survey of the Road from Briery to the Magazine ...
...the Overseer of the Road from Little Roanoak Church to the Magazine ...
* * *
...the Surveyor of the Road from the head of Mountain Creek to Ralph Sheltons ...

4 November 1760, Page 190
Joseph Morton, Dudly Barksdale, and John Pettus are appointed to Proportion the Hands to work on the Road from the Magazine, to Little Roanoak Bridge, and on the Road leading from Ruffins Quarter to Little Roanoak Bridge, and on the Road leading from Prince Edward Line to Little Roanoak Bridge

4 November 1760, Page 190
John Hearndon is appointed Surveyor of the road leading from Roysters ferry to the Court House Church, And it is ordered that he together with the usual assistance that worked on the said Road, do forth with Clear & keep the same in Repair According to Law.

4 November 1760, Page 190
On the Petition of the Inhabitants of this County for a Road to be laid off and Cleared the best and most Convenient Way by Grayors out of the Road by Richard Wittons Gent^l, It is ordered that Joseph Williams, John Cox, Thomas Pettus, and William Goode or any three of them being first Sworn &c, do diligently View and Examine the way where Such Road is Petitioned for, and Report to the next Court the Conveniency &c

4 November 1760, Page 197
William Caldwell and David Caldwell Gent^l are appointed to Regulate the Hands to work under, Richard Dudgeons and Thomas Boughon.

4 November 1760, Page 192
On the Petition of Thomas Lanier Gent^l. leave is granted him to Clear a Road from William Hills to Allens Creek, and it is ordered that he together with the following Male Labouring Tithables, to wit, his male Labouring Tithable Persons, William Hill's, Harwood Jones, Tignal Jones, and Vinenal Jones hans, do forth with lay open Clear & Keep the same in Repair according to Law.

2 December 1760, Page 201
Order'd that the Present Collector Pay Samuel Comer for the Building of Two Bridges over Cubb and Louse Creek, also another over Wards fork When finished, out of the Depositum in his Hands

2 December 1760, Page 201
Ordered that James Taylor Pay Francis Harwood one Pound Ten shillings Current Money, for the Maintenance of a Bridge over Wards fork Creek.

2 December 1760, Page 201
Order'd that the Collector Pay Richard Witton and Thomas Tabb Gentlemen what Money he hath in his Hands, for the Building of a Bridge over Meherrin River.

2 December 1760, Page 201
Order'd that a Bridge be Erected and Built over Little Roanake River at or near the Place where the old Bridge now Stands near the Plantation of James Taylor & Clement Read Gent^l. And Thomas Bouldin,

James Taylor, and Thomas Bedford Gent^l. are appointed and Desired to Lett the same to undertakers upon Such Terms, and to be built in Such Manner as they shall think fit & Convenient

2 December 1760, Page 201
Order'd that Henry Isbell be Summoned to appear at the next Court to be held for this County, to answer a Certain Complaint Existed against him, Concerning a Bridge over Little Roanoak River

2 December 1760, Page 201
Order'd that the Road whereof George Phillups was late Surveyor from Cox's road towards Stoke Bridge, be Continued, and Thomas Chambers is appointed Surveyor thereof, and it is ordered that he together with the assistance that assisted the said George on the said road, do forthwith Clear & Keep the same in Repair According to Law.

2 December 1760, Page 201
Nathaniel Christian is appointed Surveyor of the Road leading from the said Christians to Bedford County Line, And it is ordered that he together with the male Labouring Tithable Persons Convenient thereto do forthwith Clear & Keep the same in Repair According Law.

2 December 1760, Page 202
James Williams is appointed Surveyor of the Road Whereof George Vaughn was late Surveyor, And it is ordered that he together with the assistance that assisted the said George do forthwith Clear & keep the same in Repair According to Law.

2 December 1760, Page 202
James Williams is appointed Surveyor of the Road from Three Mile tree to the Head of the Haw Branch to the Cart Path that Leads from Callehams Road to the Chappell, and it is ordered that he together with the following Assistance, towit, George Vaughn Samuel Kirk, James Thomason, James Campbell, Abraham Merriman, Epharim Bowing, W^m. Shorte Henry Cavus, W^m. Slaughter, William Bowing Senr. Robert Bowing, Jessee Brown, William Bowing junior, David Bowing Isaac Johnson, and James Williams, do forthwith Clear and keep the same in Repair According to Law.

2 December 1760, Page 207
John Fuquay and Joseph Bayse two of the Persons appointed by an order of the last Court to View and Examine a way where a Road was Petitioned for leading up the hill from Bookers ferry, this day Returned a Report thereon, And thereupon the same is Established a Road and the said Joseph Bayse is appointed Surveyor thereof And, James Hunt Gent[l] are appointed and desired to proportion the Hands to work on the said Road.

2 December 1760, Page 208
Daniel Wynn is appointed Surveyor of the Road Leading from Reedy Creek Church to Daniel Hayse, and it is ordered that he together with the Hands Convenient thereto, do forthwith Cleark and keep the same in Repair According to Law.

2 December 1760, Page 211
William Poole is appointed Surveyor of the Road whereof Richard Coleman was late Surveyor, And it is ordered that he together with the Assistance that Assisted the said Richard on the said Road, do forthwith Clear and keep the same in Repair according to Law.

3 December 1760, Page 215
David Caldwell who Stands Presented by the Grand jury for not keeping Cubb creek Bridge in Repair According to Law. This day appeared, and on hearing his Reasons, It is ordered that the said Presentment be Dismissed

3 December 1760, Page 216
On the Petition of Richard Witton Gent[l] leave is granted him to keep the road open leading by his House to Gills ordinary

3 December 1760, Page 216
Mathew Marable, and Joseph Williams Gentlemen, are appointed and Desired to Lett the Repairing of the Bridge by John Cox's to undertakers upon Such Terms as they think fit & Convenient

3 February 1761, Page 219
It is ordered that Richard Witton, Thomas Lanier, John Thompson, and John Humphris or any three of them being first Sworn before a Majistrate of this County, do diligently View and Examine a way from

John Humphris to Saffolds foard on Meherrin River, and from thence to flat Rock by Trumans Mill, and Report to the next Court the Conveniency &c

3 March 1761, Page 221
On the Petition of John Ragsdale leave is granted him to make a Bridle way leading from the Ready Creek Church to flat Rock Church, and Cornelius Priest, John Sanford & Baxter Ragsdale are appointed and Desired to direct the way where Such Road is intended to be made.

3 March 1761, Page 221
Edward Colbreath, Peter Colbreath, and Samuel Wilson the Persons appointed by the Last Court to View and Examine a way where Road was Petitioned for, leading from Grassey Creek to Roysters Ferry, this day Returnd a Report thereon, and thereupon the same is Established a Road, and William Royster, Richard Yancey, & Edward Colbreath are appointed to nominate Persons to Work on the said Road

3 March 1761, Page 223
John Foster is appointed Surveyor of the Road whereof Gabriel Toombs was Late Surveyor, And it is Ordered that he together with the Assistance that Mackerness Goode shall appoint, do forthwith Clear & keep the same in Repair According to Law.

4 March 1761, Page 232
Lyddal Bacon, and Thomas Tabb Gent[l] are appointed and desired to View and Examine the Bridges built over Meherrin River by Henry Isbell, and Report to the next Court, And it is ordered that the said Isbell be Summoned to appear at the next Court to answer Such things as shall be objected against of and Concerning the said Bridges

7 April 1761, Page 252
Mathew Williams & William Johnson who being formerly appointed by an order of this Court to View a way for a Road from the Round Meadow into flat rock road, this day Returned a Report thereon which is ordered to be Disolved

7 April 1761, Page 254
Edmund Toombs is appointed Surveyor of the Road whereof John Foster was late Surveyor, And it is ordered that he together with the usual Assistance do forthwith Clear and keep the same in Repair According to Law.

7 April 1761, Page 259
Richard Naul is appointed Surveyor of the Road whereof Francis Clarke was late Surveyor, and it is ordered that he together with the usual Assistance do forthwith Clear & Keep the same in Repair according to law.

7 April 1761, Page 260
On the Petition of John Owen leave is granted him to make a Bridle way from his House into Stewarts Ferry Road

7 April 1761, Page 260
Nathan Austin, Alexander Joyce and William Holt or any two of them being first Sworn &c, are ordered to View and Marke out a road that was formerly ordered to be Viewed by Richard Austin and the said Joyce, and Report to the next Court the Conveniency or Inconveniency thereof

7 April 1761, Page 261
Edmund Beard is appointed Survey of the Road whereof Samuel Johnson was late Surveyor, And it is ordered that he together with the assistance that assisted the said Samuel on the said Road do forthwith Clear and keep the same in Repair According to Law

5 May 1761, Page 3
Grand Jury Presentments
...the Surveyor of the Road from Blanks's Ferry to the Parsons Barn...
* * *

...the Surveyor of the Road from Kings Road to Wades Ferry ...

...the Surveyor of the Road called Kings road...
...the Surveyor of Willinghams Road ...
...the Surveyor Cocks Road...

5 May 1761, Page 7
John Scott is appointed Surveyor of the Road whereof William Petty Pool was late Surveyor, and it is ordered that he together with the Hands that worked on the said Road, do forthwith Clear & Keep the same in Repair According to Law.

5 May 1761, Page 8
Isaac Johnson is appointed Surveyor of the Road whereof Henry Williams was late Surveyor, and it is ordered that he together with the usual assistance that worked on the said Road, do forthwith Clear & keep the same in Repair according to Law.

5 May 1761, Page 8
John Clarke is appointed Surveyor of the Road Whereof Stephen Mallet was late Surveyor, It is ordered that the said Clarke together with the assistance that worked on the said Road under the aforesaid Mallet do forthwith Clear & keep the same in Repair According to Law.

5 May 1761, Page 17
Charles Yancey is appointed Surveyor of the Road Whereof Samuel Morton was late Surveyor, And it is ordered that he together with the Hands that worked under the said Morton do forthwith Clear & keep the same in Repair According to Law.

6 May 1761, Page 20
On the Petition of George Foster and others, for a Road to be laid off and Cleard, the best and most Convenientest Way from Jones's Road into the Irish Road by George Fosters, It is ordered that Francis Petty, George Wells, Thomas Hatchell, Daniel Hankins & George Moore or any two of them being first Sworn &c do dilingenly View and Examine the way where such Road is Petitioned for, and Report to the next Court the Conveniency &c.

6 May 1761, Page 30
Joel Chandler is appointed Surveyor of the Road from Coxes to the Middle Meherrin, And it is ordered that he together with all the Male Labouring Tithable Persons Convenient thereto do forthwith assist him in Clearing and keeping the same in Repair According to Law.

2 June 1761, Page 46
Ordered that a Bridge be Erected and built over flat rocke at or near the old ford, and David Garland and John Jennings Gentlemen are appointed and desired to let the same to undertakers for Seven Years, to be built in Such manner and form and upon Such Terms as they in their Discretion shall think fit & Convenient.

2 June 1761, Page 47
Bryant Cocker is appointed Surveyor of the Road whereof William Drew was late Surveyor. And it is ordered that he together with the assistance that assisted the said William on the said Road do forthwith Clear and keep the same in Repair According to Law.

2 June 1761, Page 47
Joseph Bayse is appointed Surveyor of the road whereof Charles Hunt was late Surveyor, And it is ordered that he together with the assistance that assisted the said Hunt on the said Road, do forthwith Clear and keep the same in Repair According to Law.

2 June 1761, Page 47
Ordered that Thomas Lanier, Thomas Moore and Tignall Jones or any two of them being first Sworn &c do diligently view and Examine a way for a road from Butchers Creek by John Potters to Allens Creek Bridge and Report to the next Court the Conveniency thereof &c.

2 June 1761, Page 49
David Halliburton is appointed Surveyor of the Road whereof John Hearndon was late Surveyor, And it is ordered that the said David together with the assistance that assisted the said John on the said Road, do forthwith Clear and keep the same in Repair according to Law.

2 June 1761, Page 50
Francis Petty, George Wells and Thomas Hatchell three of the Persons appointed by an order of the last Court to view and Examine the way

where a Road was Petitioned for by George Foster leading from Jones Road into the Irish road by George Fosters, This day returned a report thereon, and Thereupon the same is Established a Road. And George Foster is appointd Surveyor thereof and it is ordered that he together with his own Male labouring Tithable Persons, James Foster & his, Thomas Hatchill, David Roberts, & William Foster do forthwith lay open Clear and Keep the same in Repair according to Law.

2 June 1761, Page 50
On the Petition of John Speed Gentleman for a Road to be laid off and Cleared the best and most Convenient way from the said Speeds into Mizes foard Road, And it is ordered that George Jefferson and Thomas Lanier Gentl do view and Examine the way where Such road is Petitioned for & Report to the next Court the Conveniency or Inconveniency thereof.

2 June 1761, Page 52
On the Petition of Thomas Harvey & others for a road to be laid off and Cleared the best and most Convenient Way from Bedford line to Randolphs road It is ordered that Mathew Watson, Elisha White and John Pallet, do diligently View & examine the way where such road is Petitioned for and Report to the next Court the Conveniency or inconveniency thereof

3 June 1761, Page 67
Samuel Johnson is appointed Surveyor of the Road Whereof Edmund Beard was late Surveyor, and it is ordered that he together with the assistance that assisted the said Beard on the said Road, do forth with Clear & keep the same in Repair According to Law.

3 June 1761, Page 67
James Taylor Gent[1] is appointed to Nominate and Proportion the Hands to work on the Roads whereof Clement Read and Samuel Johnson is Surveyors.

3 June 1761, Page 67
John White is appointed Surveyor of the Road whereof Thomas Wynn was late Surveyor, And it is ordered that he together with the assistance that assisted the said Thomas on the said Road do forthwith Clear and keep the same in Repair According to Law.

3 June 1761, Page 67
Thomas Hix is appointed Surveyor of the Road Whereof Henry Williams was late Surveyor, And it is ordered that he together with the Assistance that assisted the said Henry on the said Road do forthwith Clear and Keep the same in Repair According to Law

7 July 1761, Page 72
On the Petition of Henry Isbell, leave is given him untill the Last day of September next to Compleat the Bridges over Meherrin River

7 July 1761, Page 73
Edward Atkins is appointed Surveyor of the Road whereof Samuel Johnson was late Surveyor, and it is ordered that he together with the assistance that assisted the said Johnson the said Road do forthwith Clear & keep the same in Repair according to Law.

7 July 1761, Page 73
Mathew Turner is appointed Surveyor of the Road whereof Richard Palmer was late Surveyor, and it is ordered that he together with the same Hands that assisted the said Palmer on the said Road do forth with Clear and keep the same in Repair According to Law.

7 July 1761, Page 74
It is ordered that Robert Caldwell, David George and David Logan (or any two of them being first Sworn &c) do View & Examine a way for a Road from Louse Creek Bridge to Fuquays Ferry & report to the next Court the Conveniency or inconveniency thereof.

7 July 1761, Page 82
Thomas Moore and Thomas Lanier two of the Persons appointed by an order of the Last Court to View and Examine the way where a Road was Petitioned for, the best and most Convenient way from John Potters to Allens Creek Bridge, this day Returned a Report thereon, And thereupon the same is Established a Road, and Feild Farrar is appointed Surveyor thereof, and it is ordered that he together with Such hands as Thomas Lanier Gent[l] shall Nominate do forth with Clear and keep the same in Repair According to Law.

7 July 1761, Page 82
On the Petition of George Estis & others for a Road to be laid off and Cleared, the best and most Convenient way from Cocks Quarter on

Roanoak to Kings Road, And thereupon it is Established a Road, and the said Estis is appointed Surveyor thereof, and it is ordered that he together with the following Male Labouring Tithable Persons, to wit, Randolphs Hands, Henry Mayse, Cocks's John Hite, Daniel Mayse, and John Goods Hands do forthwith lay open Clear & keep the same in Repair According to Law.

4 August 1761, Page 86
Order'd that Reps Jones, Robert Liveritt and John Ussery, or any two of them being first Sworn &c, do diligently View & Examine the Way from Hust Brides to Flat Rock Church and Report to the next Court the Conveniency &c.

4 August 1761, Page 88
Robert Caldwell and David George two of the Persons appointed by an order of this Court to view and Examine a way for a Road from Louse Creek Bridge to Fuquays Ferry, This day Returned a Report thereon, and thereupon the same is Established a Road and the Surveyor of the former Road is appointed Surveyor thereof, and it is ordered that he together with the Hands that Worked on Dudgeons Road, do forthwith Clear & keep the same in Repair According to Law.

4 August 1761, Page 88
John Speed Gentlemen who was appointed by a former order of this Court to View and Examine a way for a Road that was in Dispute Between Robert Langley and Robert Alexander, This day Returned his Report thereon. Thereupon it is ordered that the old Road be Continued, and that the new one be Discontinued

4 August 1761, Page 89
Mathew Watson, Elisha White, and John Pallet who were appointed by an order of the last Court to view & Examine the way where a Road was Petitioned for by Thomas Harvey leading from Bedford line to Randolphs Road, This day Returned a Report thereon. And thereupon the same is Established a Road and John Harvey is appointed Surveyor thereof, and he together with James Gideon, William Owle, John Wood, Andersons Hands Nathaniel Christians William Harvey, Samuel Cothmans, John Harveys, Lawrance Bayers, and Benjamin Raden, do forthwith Clear & keep the same in Repair According to Law.

4 August 1761, Page 92
William Goode Gent^1. is appointed and Desired to Regulate the Hands & appoint Surveyors of the Roads from Blanks Ferry to Fosters Ordinary, as he shall think fit & most Convenient.

1 September 1761, Page 107
Thomas Chambers and Charles Parrish being two of the Persons appointed an order of this Court, to View and Examine a way for a Road to Flat rock Church, this day Returned a Report thereon, and the same is Established a Road, Edward Waller is appointed Surveyor thereof, and it is ordered that he together with the Hands Convenient thereto, do forthwith Clear and Keep the same in Repair According to Law.

1 September 1761, Page 108
Robert Beasley, Daniel Claiborne & Jarrel Willingham or any Two of them (being first Sworn &c) are ordered to View & Examine a way for a Road from the North Meherrin to Reedy Creek Church and Report to the next Court the Conveniency &c.

1 September 1761, Page 111
Robert Weakley and others who were appointed by a former order of this Court to View & Examine a way for a Road from Hunts Road to Grymes's Foard in Prince Edward line, this day Returned a Report thereon, there upon the same is Established a Road, and James Hunt and David Caldwell Gentlemen are appointed & desired to Nominate an overseer for the said Road & Hands to work under him, And it is ordered that he together with such Hands as they shall appoint do forth with Clear & keep the same in Repair According to Law.

2 September 1761, Page 120
Richard Naul is appointed Surveyor of the Road whereof Francis Clarke was late Surveyor, And It is ordered that he together with the assistance that assisted the said Clarke on the said Road, do fortwith Clear and keep the same in Repair According to Law.

3 September 1761, Page 138
Richard Williams is appointed Surveyor of the Road from Daniel Mayse to the old Trap, And It is ordered that he together with the following Male Labouring Tithable Persons, to wit, Richard Williams, Mathew

Burt, John Hazlewood, John White Isaac R   , William Chiswell, David Hopkins, do forth with lay open Clear and keep the same in Repair According to Law.

3 September 1761, Page 138
Henry Crenshaw is appointed Surveyor of the Road from the old Trap to Reedy Creek Church, And It is ordered that he together with the following Male Labouring Tithable Persons, to wit, John Strann, John White, John Smith, Jeffery Murrel, John Morgan, Epharim Pucket, Benjamin Pollard, Daniel Wynn & John Ross do forth with Clear and keep the same in Repair According to Law.

3 September 1761, Page 141
George Jefferson, John Earl, & William Robinson or any two of them (being first Sworn &c.) are ordered to View & Examine a Way for a Road from George Jeffersons Store, by a John Speeds to Cozens's, and Report to the next Court the Conveniency or Inconveniency thereof.

6 October 1761, Page 144
Thomas Rogers, Tarlton East, Peter Franklin and John Smith, or any two of them being first Sworn &c, are ordered to View & Examine a way for a Road from Turnip Creek Bridge to the new Church, and Report to the next Court the Conveniency or inconveniency thereof.

6 October 1761, Page 144
On the Petition of Thomas Pound leave is granted him to make a Bridle way from Thomas Pound's up Ledbetter Creek to the road.

6 October 1761, Page 144
Charles Stoke and John Thornton, being first Sworn &c, are ordered to View and Examine a way for a Road from Dry Creek to Reedy Creek Church and Report to the next Court the Conveniency or Inconveniency &c.

7 October 1761, Page 157
Joseph Fuquay is appointed Surveyor of the Road from Fuquays ferry to Bookers Ferry Road, And it is ordered that he together with the Hands Convenient thereto, do forthwith Clear and keep the same in Repair According to Law.

2 November 1761, Page 175
Order'd that a Bridge be Erected & built over Little Roanoke River at or near a Place Called the Mossing foard, And Thomas Bouldin & Thomas Bedford Gentlemen are appointed to lett the same to undertakers for Seven Years and to be built in such manner and form as they in their Discretion shall think fit and Convenient.

2 November 1761, Page 175
Lydal Bacon, & Thomas Tabb Gentlemen are appointed to Proportion the Hands to work on the Several Roads whereof Thomas Wynn, John Hix, Henry Blagrave, and John Scott are Surveyors.

2 November 1761, Page 175
William Robinson is appointed Surveyor of the Road where of Stephen Evans was late Surveyor. And Mathew Marable Gentleman is appointed to Proportion the Hands to work on the said Road, Cox's Road and the Road that leads from this Court House to the Parsons Barn, And it is ordered that the Surveyors of the Several Roads together with such Hands as the said Marable shall appoint, do forthwith Clear and keep the same in Repair According to Law.

2 November 1761, Page 175
Daniel May is appointed Surveyor of the Road whereof George Estis was late Surveyor, And it is ordered that he together with the Assistance that assisted the said Estis on the said Road, do forthwith Clear and keep the same in Repair According to Law.

2 November 1761, Page 178
Jacob Royster is appointed Surveyor of the Road leading from Camps Ferry to Cross Road from Roysters ferry to Colbreaths, And it is ordered tht he together with the Hands Convenient thereto do forthwith Clear & keep the same in Repair According to Law.

2 November 1761, Page 178
John Jones is appointed Surveyor of the Road leading from the mouth of Aarons Creek to the Cross Road from Roysters ferry to Colbreaths And it is ordered that he together with the Hands Convenient thereto do forthwith Clear & keep the same in Repair According to Law.

2 November 1761, Page 179
Grand Jury Presentments
...the overseer of the Road from the Magazine to Little Roanoke Church ...
...the Overseer of the Road from the fork of Randolphs Road to the Parsons Barn...
* * *
...the Overseer of the Road from Wilmoths to Cox's ...

2 November 1761, Page 180
John Hobson is appointed Surveyor of the Road Whereof John Hawkins was late Surveyor, And it is Ordered that he together with the Assistance that assisted the said Hawkins on the said Road doth forthwith Clear & keep the same in Repair According to Law.

1 December 1761, Page 181
County Levy
To Thomas Bedford Gentleman for Bridge over Roanoke ...
    29.0.0
* * *
To Ralph Shelton the Proportionable Part of this county, for Building a Bridge over Notoway River ...
    5.6.4

1 December 1761, Page 185
Henry Deloney and John Speed Gentlemen are appointed to View and Examine the best and most Convenient way for a Road from the Old School House into Inghrams Road, and Report to the next Court the Conveniency or Inconveniency thereof.

1 December 1761, Page 194
Mathew Marable and William Goode Gentleman, are appointed to view and marke out a way for a Road the best and most Convenient way around the plantation of Robert Munford Gentleman, and to this Court House

1 December 1761, Page 196
John Cox Gentleman came into Court and undertook to keep the Bridge over Meherrin at his House, a Passable Bridge for Waggons &c for the Term of Seven Years from the date here of, for the Sum of Thirty Pounds Current Money, to be levi'd at the laying of the next County Levy.

2 February 1762, Page 204
David George is appointed Surveyor of the Road whereof Thomas Vaughn was late Surveyor, And it is Ordered that he together with the assistance that assisted the said Thomas doth forthwith Clear & keep the same in Repair According to Law.

2 February 1762, Page 205
David George, Robert Caldwell and Jasper Robernett or any two of them being first Sworn &c, are ordered to View the best & most Convenient way to turn the Road round David Logans Plantation & Report to the next court the Conveniency or in Conveniency thereof.

2 February 1762, Page 205
Thomas Nance is appointed Surveyor of the Road whereof Thomason was late Surveyor, And it is ordered that he together with Charles Mason & his Hands Sherwood Walton & his Hands, & the Hands that Worked under the said Thomason do forthwith Clear & keep the same in Repair According to Law.

2 February 1762, Page 205
Arther Slayton is appointed Surveyor of the Road from the Old Road to Cubb Creek, And it is ordered that he together with all the male Labouring Tithable Persons Convenient thereto, do forthwith lay open Clear & keep the same in Repair According to Law.

2 February 1762, Page 205
Robert Weakly is appointed Surveyor of the Road from Cubb Creek half way to Prince Edward Line, And it is Ordered that he together with all the Male Labouring Tithable Persons Convenient thereto, do forthwith lay open Clear & keep the same in Repair According to Law.

2 February 1762, Page 205
Collyer Barksdale is appointed Surveyor of the Road from the half way to Prince Edward line, And it is ordered that he together with all the Male Labouring Tithable Persons Conveneint thereto, do forthwith Clear lay open & keep the same in Repair According to Law.

2 February 1762, Page 205
Reps Jones and Robert Liverett, two of the Persons appointed by a former order of this Court to View & Examine a way for a Road from Hurts Bridge to flat Rock Church, this day Return'd a Report thereon, thereupon the same is Established a Road, and Reps Jones is appointed Surveyor thereof, and it is Ordered that he together with the male Labouring Tithable Persons Convenient thereto, do forthwith Lay open Clear & keep the same in Repair According to Law.

2 February 1762, Page 205
William Davis, William Harvey, & Duglass Watson or any two of them being first Sworn &c are ordered to View & Examine the best & most Convenient Way for a Road from the head of Bear Creek into Randolphs Road, & Report to the next Court the Conveniency or in Conveniency thereof.

3 February 1762, Page 218
On the Petition of David Christopher leave is granted him to Turn the Road Round his Plantation, that leads from this Court House to Humphris's Ordinary.

3 February 1762, Page 218
Peter Jefferson is appointed Surveyor of the Road Whereof George Farrar was late Surveyor, And it is ordered that he together with the assistance that assisted the said Farrar on the said Road doth forthwith Clear & keep the same in Repair According to Law.

3 February 1762, Page 218
John Cargill is appointed Surveyor of the Road whereof William Drew was late Surveyor, And it is ordered that he together with the Hands that worked on the said Road under the said Drew doth forthwith Clear & keep the same in Repair According to Law.

3 March 1762, Page 219
Duglass Watson & William Davis being two of the Persons appointed by a former order of this Court to view & Examine the best and & most Convenient way for a Road from the Head of Bear Creek into Randolph's Road, This day Returned a Report thereon, And thereupon the same is Established a Road & William Davis is appointed Surveyor there of, and it is ordered that he together with the Male Labouring Tithable Persons Convenient thereto, doth forthwith Clear & keep the same in Repair according to Law.

2 March 1762, Page 223
Ordered that John Clarke, Thomas Adams and Batt Crowder or any Two of them being first Sworn &c do view and Examine the best and most Convenient way for a Road from Ellidges Old Rase Paths to Saffolds ford, and Report to the next Court the Conveniency or in Conveniency thereof.

2 March 1762, Page 223
Ordered that Thomas Spencer, George Foster and George Foster jun[r]. or any Two of them being first Sworn &c, do diligently view & Examine the best most Convenient way for a Road from Godfry Jones's Road into Bouldins Road, and Report to the next Court the Conveniency or in Conveniency thereof.

6 April 1762, Page 2
George Jefferson is appointed Surveyor of the Road whereof George Baskervill was late Surveyor

6 April 1762, Page 3
Mathew Watson & James Hunt Gen[t] are appointed to view a way for a Road According to Thomas Harveys Petition, and make Report thereof to the next Court

7 April 1762, Page 5
William Hill is appointed Surveyor of the Road Whereof [part of page missing] Dun was late Surveyor

7 April 1762, Page 12
Lydal Bacon Genl is appointed by this Court to Join with the Gentlemen appointed by Amelia Court to let the Building of a Bridge over Notway River by Hampton Wades.

4 May 1762, Page 26
Charles Allen is appointed Surveyor of the Road from Wade Ferry to Kings Road and it is order'd that he together with the Hands Convenient thereto do forth with Clear and keep the same in Repair according to Law.

4 May 1762, Page 27
Robert Larke is appointed Surveyor of the Road Whereof Saml Holmes was late Surveyor.

4 May 1762, Page 27
Thomas Leigon is appointed Surveyor of the Road Whereof David Gwin was late Surveyor.

4 May 1762, Page 27
John Wynn is appointed Surveyor of the Road from McConnicos to Thomas Wynns Old Muster field, and Lyddal Bacon Gentleman is appointed to Regulate the Hands &c

4 May 1762, Page 27
Thomas Nance is appointed Surveyor of the Road from Kings Road to Willinghams Road.

4 May 1762, Page 27
Ordered that Thomas Spencer, George Foster Senr. George Foster Junr. and James Foster do view the best and most Convenient way for a Road from the Irish Road to the new Church, and Report to the next Court the Conveniency &c

4 May 1762, Page 27
Rober Chappell is appointed Surveyor of the Road whereof Daniel Hay was late Surveyor.

1 June 1762, Page 32
Nicholas Maynard is appointed Surveyor of the Road whereof John Brassey was late Surveyor.

2 June 1762, Page 42
Henry Deloney, John Speed and Edmund Taylor Gen^t or any two of them are appointed to view and Examine the best and most Convenient Place for a Bride over Meherrin River near Mises Ford and Report to the next Court the Conveniency &c

6 July 1762, Page 55
John Ragsdale, Thomas Edwards, and Jessee Ozling or any two of them being first Sworn &c are appointed to view the best and most Convenient Way for a Road from the Glebe of Cumberland Parish to flat Rock and Reedy Creek Churches, and Report to the next Court the Conveniency &c

6 July 1762, Page 56
Joseph Fuquay is appointed Surveyor of the Road from Fuquays Ferry to Louse Creek Bridge, And it is ordered that he with the hands Convenient thereto do forthwith Clear and keep the same in Repair according to Law.

6 July 1762, Page 58
Ordered that James Hunt, James Taylor, David Caldwell, and Elisha White Gent are appointed to Regulate the Hands and Roads under the Several Overseers from Fuquays to Wards fork Bridge and from thence to Prince Edward Line and upwards as they shall think most Proper.

6 July 1762, Page 58
Robert Beasly is appointed Surveyor of the Road from the Miery Branch to Willinghams Road, and Lydal Bacon Gent is appointed to Proportion the Hands to Work under the said Beasly.

3 August 1762, Page 77
Tarlton East and John Smith two of the Persons appointed by an Order to this Court to View and Examine the best and most Convenient Way for a Road from Turnip Creek Bridge to Cub Creek Church, this day Returned their Report thereon, thereupon the same is Established a Road.

3 August 1762, Page 79
John Wright is appointed Surveyor of the Road from Mises Ford to Brunswick Line, and it is ordered that he together with the Hands Convenient thereto, do forthwith Clear and keep the same in Repair According to Law.

5 October 1762, Page 123
Joshua Mabry &c this day returned their Report of a Road which is ordered to be Recorded

5 October 1762, Page 125
Ordered that Mathew Marable and William Goode Gent[l]. do Proportion all the hands on the Several Roads Convenient to them.

1 November 1762, Page 135
A Petition of Richard Fox and others for a Ferry ordered to be Certified

2 November 1762, Page 136
Robert Coleman is appointed Surveyor of the Road in the Room of James Coleman

2 November 1762, Page 136
William Ruffe is appointed Surveyor of the Road in the Room of William Ballard

2 November 1762, Page 137
Ordered that James Taylor and Joseph Morton Gent[n]. let the Building a Bridge over Little Roanoke River by the said Mortons for twelve Years --

2 November 1762, Page 138
Ordered that Elisha White and David Caldwell Gent[n]. let the Building a Bridge over Cubb Creek near Grymes's ford for twelve Years

2 November 1762, Page 138
Ordered that Thomas Moore Field Farrar and John Puryear do view the Road from Akins ford to Allens Creek and report to Court the Conveniency &c

2 November 1762, Page 138
John Hays is appointed Surveyor of the Road in the room of William Lydderdale

2 November 1762, Page 138
Edmund Toombs is appointed Surveyor of the Road from Phillip Jones's Path into the same Road, and it is ordered that William Baily, William Roulet, James Shelton, Phillip Jones, Robert Worsham, Daniel Handkins, Michael Gill and John Mitchell William Hatchet, John Hatchet David Roberts, James Foster and John Haney and William Fosters hands do assist in clearing the new Road --

2 November 1762, Page 138
William Foster is appointed Surveyor of the Road in the Room of George Foster

2 November 1762, Page 138
Joseph Freeman is appointed Surveyor of the Road in the Room of Richard Swepston and it is ordered that Matthew Marable and Richard Witton Gent^n. do Proportion the hands between Cox and Freeman

7 December 1762, Page 139
Ordered that George Walton John Foster and Francis Petty do View the Road from the Parish line to the Church on ash Camp and report to Court the Conveniency &c

7 December 1762, Page 140
Ordered that David Caldwell Gent do let the Building the Little Bridge below Cub Creek and the Causway &c

7 December 1762, Page 140
Ordered that David Caldwell David Logan and Seth Ward do View the road from out of Coles Road to Fuquays Ferry above Dudgeons Plantation and report to Court the Conveniency &c

8 December 1762, Page 152
Edward Mosby is appointed Surveyor of the Road in the room of Charles Allen

8 December 1762, Page 152
John Holt is appointed Surveyor of the Road in the room of Robert Weakly

8 December 1762, Page 153
Ordered that Cargills Road be turned from the Stony Hill to bleu Stone bridge the Convenient's way &c

8 December 1762, Page 153
William Hunt is appointed Surveyor of the Road from the Wood Picker to the Mittle Creek, and John Smith from Kittle Creek to the Court House and they to divide the hands &c

8 December 1762, Page 154
John Jefferis Jun[r]. is appointed Surveyor of the Road in the room of John Cox

10 February 1763, Page 1
On the Motion of Lyddal Bacon he hath leave to Continue his Roling way that leads from the said Bacons to the Main Road to be Opened at his own Expence

10 March 1763, Page 10
It is ordered that a Road be Cleared from Richard Foxes Landing to the Country Line, and the said Fox is appointed Overseer thereof and it is ordered that the Male Labouring Tithables belonging to William Davis, Baxter Davis and the Estate of Edward Davis Dec[d]., Shippe Allen Puckett, Henry King, George Keeling and Peter Williams be a gang to work on the said Road to keep the same in repair according to Law

10 March 1763, Page 12
It is ordered that John Jeffrys, Richard Swepstone, and James Coleman do View and Inspect the most Convenient way for turning the Road Round Guy Smiths Plantation and make report thereof to Court of the Conveniences and Inconveniences of the Intended Alteration

10 March 1763, Page 12
James Hunt, James Taylor, David Caldwell and Elisha White Gent[n]. are appointed to Regulate the Roads and hands in Cornwall Parish and make such Alterations as they shall think fit

10 March 1763, Page 14
Edward Mosby is appointed Overseer of the Road in the room and Precinct of Charles Allen Gent. leading from the Stony Hill to Kings Road and John Cargill and the said Charles Allen are appointed to Divide the hands

10 March 1763, Page 22
John Mitchell is appointed Overseer of the Road in the room and Precinct of Edmund Toombs

14 April 1763, Page 28
It is ordered that Edward Lewis, Zachariah Baker and John Potter do view the most Convenient way for a Road from Kemp's ferry Road over Butcher's Creek into Edmund Taylor's Road and report to Court of the Conveniencies and Inconveniencies of the same

14 April 1763, Page 29
Edward Elam is appointed Overseer of the Road in the room and Precinct of William Elam

14 April 1763, Page 29
It is ordered that John Cargill, Thomas Buckingham, John Lucas, Benjamin Atkins, Manoah Tinsley, William Johnson, Thomas Ragsdale, Daniel Cargill, Cornelius Cargill Jun[r]. with all their Male Labouring Tithables William Arrington, Abraham Brown, Robert Christopher and David Christopher do Attend William Hill in Clearing the Road from Wade's Ferry to the Court House

12 May 1763, Page 37
On the motion of Isaac Johnson for a Rolling road from his house to Johnsons Road, Charles Sullivant, Stephen Wood and Joseph Johnson are appointed Viewers &c

12 May 1763, Page 37
William Toombs is appointed Surveyor of the Road from Kings Road to Wimbish's house and John Mitchell Mich[l] Gill William Traynum George Jones John and Robert Worsham, Richard Hix Jun[r] William Foster Thomas Reads hands Daniel Hankins and his assist &c

12 May 1763, Page 38
John Naull is appointed Surveyor of the Road in the room of Richard Naull

12 May 1763, Page 42
Ordered that Nicholas Maynard William Griffin John Mangram View the most Convenients way from the road where it Crosses Island Creek and round Daniel Mitchells Plantation, and make report to the Court

12 May 1763, Page 42
William Cunningham is appointed Surveyor of the road in the room of David George

9 June 1763, page 77
Ordered that John Hite, Matthew Organ, James Dicks View and lay of the most convenientest way for a Road from John Hites to Flatt rock Road and report according to Law.

9 June 1763, Page 77
William Gee is appointed Surveyor of the Road in the room of John Lucass

9 June 1763, page 78
Ordered that Burwells hands where John Oliver lives, John Murphey, Richard Booker and the remainder of the hands that formerly worked under Richard Swepston be under Joseph Freeman

9 June 1763, Page 85
James Foster is appointed Overseer of the Road in the room of George Foster

9 June 1763, Page 85
John Speed and Benjamin Baird Gentlemen are appointed to regulate the hands under Jacob Bugg George Jefferson and the Surveyor of the Road from Miles's creek toward Maherrin

14 July 1763, Page 88
Richard Fox entered into Bond with Security according to Law for keeping a public ferry from his landing to James Blantons

14 July 1763, Page 92
Thomas Barnes is appointed Overseer of the road in the room of John Foulton from the magazine to Roanoak bridge

14 July 1763, Page 94
John Nall is appointed Overseer of the Road in the room and Presinct of Richard Nall and it is Ordered that the Male Labouring tiths that formerly Worked on the Road under the said Richard Nall and the hands belonging to Harwood Jones, Vinkler Jones, Samuel Buzbee and Alexander Fowler be a gang to keep the said Road in repair according to Law

15 July 1763, Page 118
It is Ordered that Edmund Taylor and John Camp Gentlemen do settle and Continue the most Convenient way from the Road where it crosses Island Creek as they shall think fit and return their report to Court

15 July 173, Page 118
John Satterwhite is appointed Surveyor of the Road from Matthew Marables Gentleman his House to the Court House in the room of Thomas Hix

15 July 1763, Page 118
John Hays is appointed Overseer of the Road in the room and Presinct of William Litherdale

15 July 1763, Page 119
It is Ordered that the Sheriff do pay unto Richard Ellis of Amelia County this Countys Proportion of Nine Pound Two Shillings for building a Bridge over Nottoway River

15 July 1763, Page 119
It is Ordered that the Sheriff do pay unto Ralph Shelton this Countys Proportion of five Pounds twelve Shillings and four pence for building a Bridge over Notoway River at the said Sheltons

15 July 1763, Page 119
It is Ordered that the Sheriff do pay unto John Cox twenty nine Pound for building a bridge over the South Meherrin River at this House

11 August 1763, Page 146
Ordered that the Persons returned by a List to Court, be a Gang to keep the Road in Repair whereof John Hays is Overseer

11 August 1763, Page 147
A Report being this Day returned by Edmund Taylor and John Camp Gentlemen, Ordered that a way be Established according to the said Report

11 August 1763, Page 152
A Report being this Day returned by Edward Lewis, Zachariah Baker and John Potter concerning a Road, which is Ordered to be Recorded And the said Road Established, whereof Edward Lewis is appointed Surveyor thereof, and that Edmund Taylor, John Potter, Zachariah Baker and Peter Akin Proportion the hands between the said Edward Lewis and Matthew Turnes and appoint such other hands as they shall think convenient to Assist the said Lewis

11 August 1763, Page 152
Edward Bevill is appointed Surveyor of the Road in the room and Presinct of William White

11 August 1763, Page 153
Ordered that John Camp Gentleman do Proportion the hands between Jacob and William Royster's

11 August 1763, Page 156
Ordered that the Collector do pay Peter Rawlins Nineteen Pounds fifteen Shillings his Account for building a Bridge over Cub Creek

12 August 1763, Page 212
William Hunter Appointed Overseer of the Road from the Wood Pecker Creek to the Thistle Creek

12 August 1763, Page 213
John Smith is appointed Overseer of the Road from the Thistle Creek to the Court House [?]

12 August 1763, Page 214
Richard Witton Gentleman is appointed Overseer of the Road from Joseph Greers to Mr Thomas Erskines's and that Joseph Greer, Benjamin Whitehead, John Linton, Thomas Netherly, Thomas Erskine, Phillip Poindexter and their hands be a Gang to worke upon the said Road and keep it in repair according to Law

12 August 1763, Page 218
Matthew Hay is appointed Overseer of the Road in the room and Precinct of John Hobson

8 September 1763, Page 232
Ordered that John Hays with the hands under him and Richard Ragsdale and the hands belonging to Armstead under Samuel Taylor Overseer, do clear from Colonel Harwoods ferry to the Road that goes from the Court House to Colonel Wittons

9 September 1763, Page 238
John Rogers is appointed Overseer of the Road from the Glebe in Cumberland Parish to Reedy Creek Church, and John Jenings from the said Glebe to Flat rock Church And Lyddal Bacon & David Garland Gentlemen are to Divide the hands between the said Overseers

9 September 1763, Page 253
On the Motion of John Humphries for leave to turn the Road that leads by his Plantation, Ordered that John Speed and Henry Delony Gentlemen or either of them do view the same and report to Court

13 October 1763, Page 255
James Ferrell is appointed Overseer of the Road in the room and Precinct of Jacob Bugg

13 October 1763, Page 257
James Clarke is Appointed Overseer of the Road in the room of and Precinct of Richard Ship

13 October 1763, Page 257
On the Motion of James Foster and others, it is Ordered that George Walton, Francis Petty, Frederick Nance and Elijah Wells or any three of them do view a way from Cumberland Parish line to Ash Camp Church and report to Court

13 October 1763, Page 260
John Camp Gentleman who was appointed to Proportion the hands between Jacob and William Royster, returned his Report this day which is Ordered to be Recorded

9 November 1763, Page 264
County Levy
To Robert Caldwell for Building a Bridge over a Branch of Cub Creek and maintaining of it twelve Years
$$9.10.0$$

10 November 1763, Page 270
Ordered that James Taylor Gentleman former Treasurer of this County do pay Josiah Morton fourteen Pounds fourteen Shillings out of the Money in his hands for, for building a Bridge over little Roan Oak

10 November 1763, Page 272
Grand Jury Presentments
We Present the Surveyor of the Road from Taylors Ferry to Clarks Ordinary for not keeping Posts of directions to the same.
We Present the Surveyor of the Road from this Court House to John Glasses for not keeping the same in repair and Sign Posts on the same.
We Present the Surveyor of the Road from Cargills Road to Cox's Road for not keeping the same in repair and Sign Posts on the same.
We present the several Surveyors of the Road from Wilmoths Ordinary to Bollings Bridge and likewise up the same to George Pattillo's and also of the Road over Mortons Bridge to the Magazine for not keeping the same in repair and Sign Posts thereon according to Law.
Likewise from Nalls Shop to Pankeys Store for not keeping the same in repair and Sign Posts thereon according to Law.

We Present the Several Surveyors of the Road from Erskins Store to great Nottoway Bridge for not keeping the same in repair and also for not keeping Sign Posts thereon according to Law.

10 November 1763, Page 273
Ordered that Samuel Young the Overseer of the Road in the room and Precinct of Samuel Bugg

10 November 1763, Page 273
Ordered that John Camp, Peter Akin and Zachariah Baker do Proportion the hands between Matthew Turner and Edward Lewis

10 November 1763, Page 274
A Report for a Road that was Motioned for by James Foster and others, being this Day returned and Ordered to be Recorded And that all the hands between Willinghams Road and the Road to Ash Camp Church to the Parish Line do open and clear the same according to Law. And Elijah Wells is appointed Overseer of the same

10 November 1763, Page 275
Hezekiah Jackson is appointed Overseer of the Road from Bryery to Bollings Road and Ordered that the hands of William Watkins, George Jones, Samuel Wimbush, and James Shelton be a Gang to keep the same in repair according to Law

10 November 1763, Page 275
Ordered that the hands of George Moore, Peter Beasley and James Foster be added to the gang appointed to keep he Road in repair Whereof James Clarke is Overseer

10 November 1763, Page 276
Sherwood Walton is appointed Overseer of the Road in the Room and Precinct of Thomas Nance

11 November 1763, Page 285
Ordered that Clement Read do appoint the hands to Work upon Fosters Road

8 December 1763, Page 298
John Smith is appointed Overseer of the Road in the room and Precinct of Stephen Wood

8 December 1763, Page 298
Benjamin Clarke is appointed Overseer of the Road from Coxes to the Middle Maherrin River

8 December 1763, Page 298
A Report was this day returned by John Camp Gentleman and Peter Akin which is Ordered to be Recorded.

8 December 1763, Page 298
Peter Akin is appointed Overseer of the Road from Camps Ferry to the fork of the Road to the School House.

8 December 1763, Page 299
A Report being this day returned by John Hight and others was Ordered to be Recorded

8 December 1763, Page 299
John Hight is appointed Overseer of the Road aforesaid and David Garland is appointed to Proportion the hands between the said Hight, John Calliham and James Williams

8 December 1763, Page 299
Ordered that Joseph Williams and Henry Blagrave Gentlemen do Proportion the hands upon Johnsons and Waltons Road, between John Smith, William Rivers and Isaac Johnson

8 December 1763, Page 300
On the Motion of Thomas Bedford Gentleman he has leave to open a Bridle way from his Mill to Martins Roling Road, and that the said Bedford have the directing of it And that Nehemiah Frank, Robert Davis, John Sullivant, Joshua Wharton, William Elliss, Richard Elliss, Benjamin Smith, Edward Harris and Maynard Harris do Assist in clearing the said intended way -- but not to be Exhivitted from Working on other Roads

8 December 1763, Page 300
Ordered that David Garland and Thomas Tabb Gentlemen do let the Building of the Bridge a cross Flat rock Creek by Captain John Jinings's to the lowest Bidder according to Law

8 December 1763, Page 307
Ordered that Matthew Goode Thomas Neal & John Cox Junior or any two of them do view the most convenient Way for a Road Round Robert Munford Gentleman's Plantation at Finney Wood and with the said Munfords hands open accordingly

9 February 1764, Page 309
On the Motion of Edward Ealam for leave to turn the Road round his Plantation, Ordered that William Goode Mackness Goode and Reps Osborne do view the same and make report to Court

9 February 1764, Page 309
A Report of Charles Sullivant, Stephen Wood and Joseph Johnson concerning a Roling Road Petitioned for by Isaac Johnson was this Day returned and ordered to be Recorded

9 February 1764, Page 311
John Mitchell is appointed Overseer of the Road in the room and Precinct of William Toombs

9 February 1764, Page 313
On the Motion of the Inhabitants of Cornwell Parrish, it is Ordered that Thomas Spencer, John Haney, Richard Dabbs and Clement Read do view the most Convenientest way for a Road from the Magazine to Fosters Road and report to Court

9 February 1764, Page 313
Merritt Bland is appointed Overseer of the Road from Kotnops [?] Road to Allens Creek by the new Church

9 February 1764, Page 314
Ordered that Matthew Watson, John White and Elisha White are appointed to Proportion the hands on the Road whereof Arther Slaton is Surveyor between the said Arthur Slayton, Francis Mann and Peter Rawlins, also the Road whereof Collier Barksdale was Overseer between John Holt,

William Price, Samuel Allen and the said Collier Barksdale, and also to view a way for a Road for Wards fork Bridge to the new Bridge on Cubb Creek and also from the said new Bridge to Alexander Joyces Road

8 March 1764, Page 316
On the Motion of Jonathan Patteson, leave is granted him to turn the Road round his Plantation

8 March 1764, Page 316
On the Petition of Henry Blagrave Gentleman a Public Roaling way is granted him according to the said Petition, and the hands and Overseer to be according to the Prayer of the aforesaid Petition

8 March 1764, Page 319
On the Petition of Sundry Inhabitants on Crooked Creek and Adjoining thereto for a Road from Saffolds ford on Maherin River to Brunswick Line near [blank in book] Bridge, Ordered that David Garland William Taylor and Henry Vandyke do view the most Convenient way and report to Court according to Law

8 March 1764, Page 322
Robert Caldwell is appointed Overseer of the Road in the room and Precinct of James Cunningham

8 March 1764, Page 322
Daniel Gorrie is appointed Overseer of the Road from Cox's Bridge to the Middle Maherrin River Bridge and the Male Labouring Tithables that is Convenient thereto be a Gang to keep the same in repair according to Law

8 March 1764, Page 323
Lyddal Bacon Gentleman is appointed to let the Building of a Bridge over F***ing Creek by Allen Stokes's to the lower bidder according to Law

8 March 1764, Page 323
Ordered that the collector of this County do pay unto John Cox Gentleman, out of the Depositum in his hands a further allowance of twenty Shillings for building a Bridge near his House

8 March 1764, Page 323
George Walton and Francis Petty are appointed to Divide the hands between John Mitchell and Robert Breedlove

8 March 1764, Page 323
Christopher Billups Gentleman is appointed Overseer of the Road in the room and Precinct of John Knight from the head of Mountain Creek to Nottoway Bridge at Ralph Sheltons

12 April 1764, Page 332
A Report being this day returned concerning the Road that was Motioned for by Edward Ealam which was Ordered to be Recorded and the said way Established

12 April 1764, Page 338
A Report returned (on the Motion of John Humphris for leave to turn the Road) and the same was Ordered to be Recorded

10 May 1764, Page 1
Grand Jury Presentments
We Present the Surveyor of the road from Cumberland Parish line to Ash Camp Church for not keeping the same in repair according to Law,
We Present the Surveyor of the Road from the fork of Bollings Road to Willinghams Bridge ditto
We Present the Surveyor of the Road from Allens Creek to Humphries for not keeping the same in repair according to Law,

10 May 1764, Page 4
On the motion of Dennis Larke for a way to be Viewed for a road from Doctors Clack Courtneys Road to Ingrams Road upon the Ridge between Flat Creek and Dockerys Creek, It is ordered that William Lucas, Howell Colleir, Robert Larke and David Dortch or any three of them being first sworn according to Law, do View the way that the said Dennis Larke purposes for the said Road to go and report to Court of the Convenience and Inconvenience of the same --

10 May 1764, Page 4
William Jeter is appointed overseer of the road in the room of John Scott --

10 May 1764, Page 4
John Fillips is appointed overseer of the Road leading from Great Owle into the Great Road at Wilmoths ordinary and it is ordered that all the hands that worked under the former overseer be a gang to keep the said Road in repair according to Law --

10 May 1764, Page 5
On the motion of William Cocke for a road from the mouth of Butchers Creek to Church and to the Court house. It is ordered that Thomas Moore Junr. Field Farrar and Alexander Fowler being first Sworn according to Law do view the way that the said William Cocke purposes for to have the said Road to go and report to Court of the Conveniences and Inconveniences of the same --

10 May 1764, Page 7
On the motion of William Taylor for a Road to the Clerkes Office, It is ordered that David Garland, John Ragsdale and Henry Vandycke do view and lay off the most Convenients way for a road from Captain Jennings Road to the said Office and that they return their report to Court --

10 May 1764, Page 7
A report returned by Thomas Moore and others for a road ordered to be Recorded --

11 May 1764, Page 26
It is ordered that the hands of John Puryear, Joseph Rudd, Lewis Akin Henry Robertson, Thomas Moore John Roberson, John Clarke, Wm Marable, Isham Fischer [?], Conrod Messer Smith and John Kitchen be a gang to Clear the Road whereof Elisha Brooks was late appointed overseer --

12 May 1764, Page 32
Ordered that John Cox Jun. Thomas Neal, William Goode and Stephen Evans or any three of them being first sworn according to Law do view the most Convenients way for turning the Road Round Robert Munfords his Plantation at Finny Wood and report to Court according to Law

14 June 1764, Page 83
Thomas Anderson is appointed overseer of the Road in the room of John Nall --

14 June 1764, Page 83
William Brumfield is appointed overseer of the road in the room of Rees Hughes.

14 June 1764, Page 84
A Report being this day returned, for a Road which was Motioned for by William Taylor, which said Report is Ordered to be Recorded, and a way Established according thereto, and it is Ordered that all the Male Labouring Tiths belonging to Roger Atkinson, & Augustine Claiborne Gentlemen at their Quarters, David Garland's and the said William Taylors be a Gang to Clear and keep the said Road in repair according to Law

14 June 1764, Page 84
A Report for a Road being this day returned which was Motioned for by Sundry of the Inhabitants of this County which was Ordered to be Recorded and the said Road Established and Henry Vandyke is appointed Overseer from Saffolds ford to flat rock Creek and John Calliham from the said Creek to Brunswick line, Whereof David Garland and William Taylor do appointed Gangs that is Convenient to the said Road under the said Overseers to Clear and keep the said Road in repair according to Law

15 June 1764, Page 92
Ordered that the Agent for this County do pay David Garland and Thomas Tabb Gentlemen the Price due by Bond for a Bridge let by them to Isaac Brown across flat rock Creek near Capt. Jinings's

15 June 1764, Page 93
On the Petition of Sundry of the Inhabitants of Saint James Parish for a Road, Ordered that Benjamin Baird, Dennis Lark James Arnoll and Abraham Whittemore or any three of them do view the most Convenientest way according to the Prayer of the s^d. Petitioners and report to Court according to Law

15 June 1764, Page 94
Ordered that David Garland Gentleman do proportion the hands between John Jinings and John Ragsdale

12 July 1764, Page 97
On the Petition of Sundry of the Inhabitants of Saint James Parish a report was this day returned and the said Petition Ordered to be Dismissed

12 July 1764, Page 97
A Report on the Motion of Dennis Larke for a way for a Road was this day returned and Ordered to be Recorded, and the said way Established, And it is Ordered that the hands under Robert Lark and Adam Poole Surveyors of Roads do clear and keep the said Road in repair according to Law

12 July 1764, Page 98
On the Motion of Sundry of the Inhabitants of this County for turning the Road that crosses Cubb Creek at Richard Dudgeons's Plantation that leads to Fuqua's Ferry, so that the said Road goe some distance above the said Dudgeons Plantation. It is Ordered that Seth Ward, John Ward, Thomas Bouldin and James Taylor, or any three of them being first sworn as the Law directs do view and lay off the most Convenientest Way to turn the said Road in such manner as Motioned for, and report to next Court of the Conveniences and inconveniences that may attend the same

12 July 1764, Page 100
The Presentment of the Grand Jury against the several Surveyors of the Road from Erskines's Store to Great Nottoway River being returned Executed on John Ragsdale. The said Ragsdale came into Court and on hearing his Excuse The said Presentment is Ordered to be Dismissed

12 July 1764, Page 100
On the Petition of Sundry Inhabitants of this County for a Road Ordered that James Bilbo, John Humphries and Richard Swepson do view the same and report to Court according to Law

13 July 1764, Page 111
The Presentment of the Grand Jury against the Surveyor of Kings road from Martins Road to Robert Breedloves, being returned Executed on Thomas Nance, who appeared and on hearing his Excuses the said Presentment is Ordered to be Dismissed.

13 July 1764, Page 114
The Surveyor of the Road from Allens Creek to Humphris's who stands Presented by the Grand Jury for not keeping the same in repair according to Law being returned Executed and John Humphries as Surveyor thereof, who not appearing it is Considered by the Court that he for such Offence do make his fine to our Sovereign Lord the King by the Payment of fifteen Shillings Current Money for the use of the said County towards lessening the Levy thereof, and also that he pay the Costs of this Prosecution, and may be taken &c

13 July 1764, Page 120
Conrod Messer Smith is appointed Surveyor of the Road in the room and Precinct of William Mitchell

9 August 1764, Page 125
It is Ordered that the Surveyor of the Road from this Court House to M^r. Taylors ferry, do not go through the Corn field of William Dowsings but that he do goe the most Convenient way round the same

9 August 1764, Page 126
Ordered that the Agent for this County do pay unto Matthew Watson three Pounds five Shillings on Account of Money paid by John Dudgeon former Sherif to Henry Isbell James Taylor, John Ward and Seth Ward three of the Persons appointed, on the Motion of Sundry Inhabitants of this County for turning the Road from Fuquays ferry to George Pattillo's so that the said Road cross Cub Creek some small distance above Richard Dudgeon's Plantation for them to view the said way and report to Court of the Conveniencies of the same having This day returned Their report which is ordered to be Recorded and that a Road be Established according thereto, and it is Ordered that a Bridge be built over Cub Creek where The said Road is to cross, and James Taylor and David Caldwell Gentlemen are appointed to let the Building Thereof for Seven Years and the said Thornton and Caldwell are also appointed to Proportion the hands to work on the said Road, to keep The same in repair according to Law

9 August 1764, Page 126
On the Motion of Walter Coles and others they have leave to keep Open the Old road from the fork between Cub Creek and Stanton River Crossing the Old ford below Richard Dudgeons's Plantation and into the road to Pattillo's

9 August 1764, Page 126
James Barnes is appointed Surveyor of the Road in the Room of Joshua Chaves from the Horse Pen Creek to the Parsons Barn

9 August 1764, Page 127
On the Motion of James Blanton, he hath liberty to keep the ferry over Roan Oak River opposite Fox's Landing on the said River (he giving Bond and Security) Whereupon he together with sufficient Security entered into and Acknowledged Bond for that Purpose according to Law

9 August 1764, Page 129
John Johnson is appointed Surveyor of the Road in the room of Chesley Daniel from Kings Road to the Country line and it is Ordered that he have the hands of Josiah Daniel, James Kidd Miles Johnson, Joseph Michaux, John Stephens & Thomas Stephens to work on the said Road and keep it in repair according to Law

9 August 1764, Page 129
On the Petition of Charles Allen Gentleman and Sundry others setting forth that whereas the upper Inhabitants of this County on Stanton River and the Inhabitants of Hallifax do Labour under great ill conveniencies on Accountof the Road that passes from Blanks's ferry through the County to Bollings Point and as great Numbers Trade to that Place They Therefore Pray the Court to grant and Order for a Road to be opened out of the said road that crosses Blanks's ferry from William Tomasons into Willinghams Road near Mahering Bridges along the usual Tract that was formerly opened by Peter Hudson Deceased Whereupon it is Ordered that the said Road be opened and Continued according to the prayer of the said Petition agreeable to a former Order of Court directing the same, and the said William Thomason is appointed Overseer thereof

10 August 1764, Page 151
Ordered that Lyddal Bacon, Matthew Marable and Thomas Tabb Gentlemen do view the Bridge over the North Maherrin River lately repaired by

William Chandler and report to Court what they shall think the repairs done thereto by the said Chandler is Worth

13 September 1764, Page 158
Henry Hays is appointed Overseer of Road from Flatt rock Creek to Brunswick line in the room of John Calliham --

13 September 1764, Page 162
On the Motion of Abraham Maury Gentleman he has leave to have a Bridle way from his House through the Land of Obadiah Hooper and John Chandler into the Road that leads by Richard Elliotts Plantation

13 September 1764, Page 166
William Read and Abraham Martin are appointed to view and lay off the most Convenient way from the Glebe in Cornwall parish to Sandy Creek Church and that they shall appoint Overseers and hands as they shall think fit to work thereon And it is also Ordered that Thomas Bouldin and Thomas Spencer do view and lay off the most Convenient ways from the Glebe aforesaid to the several Churches in the said Parish and that they do appoint Overseers and hands as they shall think fit

11 October 1764, Page 176

Edward Crews is appointed Overseer of the Road in the Room and Precinct of John Stewart

11 October 1764, Page 176
On the Motion of Edward Colbreath for leave to turn the Road from William Roysters ferry to Carrolina line. It is Ordered that Richard Yancy, William Easley, William Perry and Peter Colbreath or any three of them being first Sworn according to Law do view the said Road and way the said Edward Colbreath Purposes to turn the same and Report to Court of the Conveniencies and inconveniencies of the intended alteration

11 October 1764, Page 177
Ordered that a Bridge be built over Flat Rock Creek where the new Road crosses the same leading from Saffolds ford to Brunwick line and Henry Vandyke and Barttelot Anderson are appointed to let the building thereof

11 October 1764, Page 177
Edward Waller is appointed Overseer of that part of the new Road leading from Saffold's ford to Brunswick line next to Henry Hayes below his Precinct

7 November 1764, Page 190
County Levy
To Matthew Watson for Money paid James Mitchell for a Bridge
    8.15.0

8 November 1764, Page 191
Samuel Peace is appointed Overseer of the Road in the Room and Precinct of Edward Waller

8 November 1764, Page 192
John Fuqua is appointed Overseer of the Road in the room and Precinct of Joseph Bayes

8 November 1764, Page 193
Ordered that Benjamin Baird, George Baskerville and Dennis Larke being first sworn according to Law, do view the two Roads leading by the Plantations of John Speed Gentleman and Sherwood Buggs and Report to Court of the Conveniencies and Inconveniencies of the same, and also which they shall think in their Opinion and Judgment is the most Convenient for the Public to be Established up and down the River that the one may be Established and the other Discontinued

8 November 1764, Page 194
Grand Jury Presentments
We Present the Keeper of Allens Creek Bridge for not keeping the said Bridge in repair according to Law
Samuel Comer for not keeping Wards Fork Bridge in Repair according to Law, in Cornwall Parish
The Overseer of the Road from Matthew Marables to the Court House, and Likewise the Overseer of the Road from the said Marables to Roanoke Church for not keeping the same in repair according to Law

The Overseer of the Road from below the Magazine at the fork to little Roanoke Church for not keeping the same in Repair according to Law

The Surveyor of the Road from Allens Creek Bridge to Taylors ferry for not keeping the same in repair according to Law in St. James's Parish

The Surveyor of the Road from Burgamy's Church to the Road that leads to Allens Creek Bridge in St. James's Parish for not keeping the same in repair according to Law

The Surveyor of the Road from Coxes Ordinary to Scotts Bridge on Maherin River in Cumberland Parish for not keeping the same in repair according to Law

The Surveyor of the Road from the Court House to Taylors Ferry on Roanoke in St. James's Parish for not keeping the same in Repair according to Law

The Surveyor of the Road from Cox's along Waltons Road to the Middle Maherrin in Cumberland Parish

The Surveyor of the Road from the middle Maherrin to Wilmoths Ordinary Cornwall Parish

The Surveyor of the Road from the Parsons Barn on Kings Road to Martins Road Cornwall Parish

8 November 1764, Page 196
Thomas Smith and Drury Moore is appointed Overseers of the Road from Barrs Element to the fork of the Reedy Creek Road in the room of John Ragsdale Gentleman. And Thomas Tabb, David Garland Gentlemen and the said Ragsdale are appointed to Divide and Proportion the hands between the said Overseers, to work on the said Road and keep the same in repair according to Law

8 November 1764, Page 196
Ordered that William Goode Gentleman Divide the hands between William Thomason, James Barnes Edward Ealam and Abraham Martin

13 December 1764, Page 198
James Sandifer is appointed Overseer of the Road in the room and Precinct of Samuel Young and it is Ordered that John Speed and Samuel Bugg Gentlemen do Proportion the hands between the said Sandifer, Jacob Mitchell and Marriott Bland

13 December 1764, Page 198
On the Motion of Wells Thompson for leave to turn the Road below his Plantation, Ordered that David Moss, John Wright and Thomas Wright being first sworn according to Law do view the said Road and the way the said Thompson Purposes to turn the same and report to Court of the Conveniencies and inconveniencies of the intended alteration

13 December 1764, Page 198
Edward Colbreath is appointed Overseer of the Road from Roysters Ferry to Carolina line in the room of William Royster

13 December 1764, Page 198
On the Motion of John Camp Gentleman for Persons to be appointed to view the most Level and Convenient way for turning the Road from his ferry to the fork of the Road. Whereupon it is Ordered that Edward Lewis, Zachariah Baker and Bozwell Wagstaff being first worn according to Law do view the said Road in the way that the said Camp Purposes to turn the same and report to Court of the Conveniencies and inconveniencies of the intended alteration

13 December 1764, Page 198
James Thompson Bardon is appointed Overseer of the Road in the room and Precinct of Edward Mosby and it is Ordered that the hands that Worked under the said Moseley Divided by Charles Allen Gentleman and John Cargill do Work under the said Bardon in clearing and keeping the said Road in repair according to Law

13 December 1764, Page 199
A Report on the Motion of William Cocke for a Road was this Day returned and Ordered to be Recorded

13 December 1764, Page 199
Richard Swepson, John Humphries and James Bilbo, the Persons appointed On the Petition of Sundry Inhabitants of this County setting forth that they lie under very great want of a Road from John Humphrises Ordinary to Saffolds ford on Maherrin River Return'd their Report which is Ordered to be Recorded in the said Road Established according thereto

13 December 1764, Page 199
Richard Yancy, William Easley, William Perry and Peter Colbreath the Persons appointed on the Motion of Edward Colbreath for leave to turn the Road from William Roysters Ferry to Carolina line, Returned their Report which is Ordered to be Recorded and the said Road Established according thereto

13 December 1764, Page 200
Ordered that John Speed Agent for this County do pay unto Henry Vandyke and Barttelot Anderson Thirteen Pounds ten Shillings it being the Price for building a Bridge let by them a Cross Flatt rock Creek to Michael Mackey

INDEX

Note: This index is arranged by subject: Personal Names; Bridges; Chapels, Churches, Glebes, Parishes; Ferries; Fords; Houses; Mills; Mountains and Hills; Ordinaries; Plantations; Quarters; Rivers, Creeks, etc.; Miscellaneous; and Roads

Personal Names

William Abbot, 50, 53
Abraham Abney, 80
Dennit Abney, 93
George Abney, 25, 62
James Adams, 104
Nathan Adams, 104
Richard Adams, 111, 129
Thomas Adams/Addams, 86, 148
William Adams, 74, 104, 122
Benjamin Adkins, 103
John Adkins, 100
John Akin, 49
Lewis Akin, 165
Peter Akin, 157, 160, 161[(2)]
Perrin Alday/Alay, 72, 104
Robert Alexander, 141
Charles Allen, 123, 127, 149, 152, 154, 169, 173
Drury Allen, 70
Samuel Allen, 163
William Allen, 27, 33
William Almond/Almon, 63, 78, 103
James Amos, 38, 73
Barttelot Anderson, 170, 174
Charles Anderson, 51
Thomas Anderson, 32, 45, 59, 91, 96, 99[(3)], 166
Anderson, 141
John Andrews, 33
Robert Andrew/Andrews, 94, 110
John Anthony, 47, 65
Mr Armstead, 126, 158
Armstead's Overseer (Samuel Taylor), 158
James Arnold/Arnoll, 8, 65, 166
William Arrington, 154
John Ashworth, 38, 40[(2)]
Leonard Ashworth, 47
Samuel Ashworth, 57
Benjamin Atkins, 74, 154

Edward Atkins, 140
John Atkins, 23, 58, 74
John Atkins's sons, 58
William Atkins, 74
Burrell Atkinson, 129
Epharim Atkinson, 129
Roger Atkinson, 166
John Austin, 23, 34, 58[(2)], 74
John Austin's sons, 58
John Austin Jur., 23
Nathan Austin, 136
Richard Austin, 87, 136
Valentine Austin, 100
Henry Avery, 126
James Avery, 126
John Avery, 126
Mathew Avery, 126
Robert Baber, 6
John Bacon, 24, 70, 74, 92, 93
Lyddall Bacon, 11, 21, 25, 32, 35, 56, 79, 80, 94, 98, 113, 114, 117, 119, 120, 127[(2)], 135, 144, 149[(2)], 150, 153, 158, 163, 169
William Bacon, 99
William Baily, 129, 152
Adam Baird, 64
Benjamin Baird, 155, 166, 171
Robert Baker, 19, 20
Zachariah Baker, 105, 119, 154, 157, 160, 173
Edmund Ballard, 128
John Ballard, 104
William Ballard, 104, 110, 151
James Thomson Barden/James Thompson Bardon, 83, 173
William Barker, 129
Collier/Collyer Barksdale, 147, 162, 163
Dudly Barksdale, 131
James Barnes, 169, 172
Thomas Barnes, 156
William Barry, 86
William Bartlete, 98
James Barton, 111
George Baskerville, 60, 63, 89, 90, 148, 171
Henry Bates, 91
John Bates, 91, 98, 111
Thomas Baughan/Baughon, 94, 123, 132
Baxter, 39
Lawrance Bayer, 141
Joseph Bayse/Bays, 82, 93, 110, 134, 138, 171

William Beal, 131
William Bean/Been, 10, 29, 129
Edmund Beard, 136, 139
John Beard, 6
Peter Beasley, 160
Robert Beasley, 142, 150
Stephen Bedford, 47, 62
Stephen Bedfords, Overseer 47
Thomas Bedford, 46, 62, 81, 106, 114[(2)], 122[(2)], 125, 133, 144, 145, 161
Captain Bedford, 41
Bedfords Overseer, 46
William Been – see William Bean
Thomas Bell, 99
William Bell, 99
Joseph Bennit, 99
Richard Berry, 74, 104
William Berry, 83
Alexander Berryhill, 111
Edward Bevill, 157
William Bevill, 49, 61
James Bilbo, 61, 167, 174
Christopher Billups, 164
Joseph Billups, 39
John Binum, 26
Mason Bishop, 129
William Black, 94, 110
Calub Blackwell, 49
Henry Blagrave, 36, 144, 161, 163
Henry Blagrave Senr., 102
Henry Blagrave Junr., 102
William Blagrave, 41
Merritt (Marriott) Bland, 162, 173
Joseph Blanks, 18, 19, 27
Richard Blanks, 104, 118, 123
James Blanton, 156, 169
Argil Blaxtone/Blaxton, 47, 71, 112
John Booker, 61
Richard Booker, 13, 93, 110, 155
William Booker, 52, 60, 61
Capt William Booker, 27
Hugh Boston, 70
Joseph Boswell/Bozwell, 69, 81, 112, 127
John Boughton, 110
Thomas Boughton, 110
Thomas Bouldin, 6, 11, 21, 24, 39, 41[(2)], 47, 48, 52, 60, 68[(2)], 76, 79, 83[(2)], 84, 102, 106[(2)], 109, 113, 114[(2)], 115, 122[(2)], 125, 132, 144, 167, 170

Capt. Bouldin, 37, 41, 46
Bouldin/Boulding, 125
David Bowing, 133
Epharim Bowing, 133
Robert Bowing, 133
William Bowing Senr., 133
William Bowing junior, 133
Thomas Bowrey, 118
John Boyd, 24, 34
John Bozeman, 98
Francis Bracy, 70, 126
John Bracy/Brassey, 91, 92, 125, 150
William Bramlet, 47
Thomas Branch, 47
Thomas Brandon, 33, 34
Francis Bray, 59
James Breedlove, 27, 44
Robert Breedlove, 164, 168
Hust Brides (perhaps Hurt's Bridge), 141
David Bridgforth, 115
Israel Bron – see Israel Brown
Elisha Brooks, 100, 106, 127, 165
Richard Brooks, 100, 106
Robert Brooks, 29, 106
William Brooks, 106
Robert Brookhouse, 30
Abraham Brown, 154
Isaac Brown, 166
Israel Brown, 27, 76
Jessee Brown, 133
John Brown, 10
Valentine Brown, 18
George Bruce, 126
James Brumfeld, 46
Willaim Brumfield, 21, 166
Thomas Buckingham, 154
Jacob Bugg, 90, 155, 158
Samuel Bugg, 105, 122, 160, 173
Sherwood Bugg, 171
David Bullock, 126
William Burgamy, 76
Daniel Burn, 63
James Burnside, 111
Mathew Burt, 143
Hutchins Burton, 31, 50, 61, 63
Burwell, 155

John Butler, 87, 94, 110
Samuel Buzbee, 156
William Byrd, 33, 64, 81, 130
Colo. Byrd, Bird, 50, 97, 105
Colo. Byrds Overseer, 50
David Caldwell, 25, 27[(2)], 48, 60, 77, 78, 80, 90, 106, 108, 111, 114, 115, 132, 134, 142, 150, 151, 152[(2)], 153, 168
Edward Caldwell, 24
James Caldwell, 32, 104
John Caldwell, 7, 18, 25, 30, 104, 111
Robert Caldwell, 111, 140, 141, 146, 159, 163
William Caldwell, 18, 30, 32, 34, 35, 37, 41, 43, 45, 48, 58, 60, 69, 76[(2)], 77, 79, 80, 96, 111, 114, 115[(2)], 116, 125, 126, 132
Nicholas Callahan/Callaham, 49
Richard Callaway, 35
William Calloway, 6, 35
David Callinghame, 129
John Callingham/Calliham, 129, 161, 166, 170
Nicholas Callingham, 129
John Camp, 127, 156, 157[(2)], 159, 160, 161, 173
Agnus Campbell, 110
James Campbell, 133
Andrew Cannaday, 74
William Carby, 43
Cornelius/Cornellius Cargill, 8, 25, 34, 36, 41, 48, 50, 53, 58, 69, 91, 96, 103, 126, 127, 128, 130
Cornelius Cargill Junr., 154
Daniel Cargill, 154
John Cargill, 8, 11, 33, 37, 50, 66, 73, 91, 96, 127, 128, 147, 154[(2)], 173
John Carrel, 61
Paul Carrington, 121, 122
Edward Carter, 89
Mathew Carter, 126
Henry Cavus, 133
Thomas Chamberland, 98
Thomas Chambers, 130, 133, 142
Joel Chandler, 60, 123, 138
John Chandler, 43, 52, 69, 112, 170
Joseph Chandler, 60
William Chandler, 128, 170
Robert Chappell, 149
Joshua Charin, 104
Thomas Charlton, 48
George Chaves/Chavus, 91, 111
Joshua Chaves, 169
Henry Childress, 96, 121, 122

Robert Childress, 5, 51
William Chiswell, 143
Gideon Chranshaw – see under Crenshaw
Nathaniel Christian, 63, 133, 141
Coris Christopher, 37
David Christopher, 147, 154
Robert Christopher, 154
Augustine Claiborne, 166
Daniel Claiborne, 70, 81, 107, 113, 127[(2)], 142
Leonard Claiborne, 45, 77[(2)]
Benjamin Clarke, 161
Francis Clarke, 125, 136, 142
James Clarke, 158, 160
John Clarke, Clark, 33, 59, 111, 137, 148, 165
John Cleaton, 98
William Cleaton, 98
Joseph Cloud, 22
Abraham Cocke/Cock/Cooke, 5, 11, 15, 25[(2)], 33, 43, 44, 57, 118
James Cocke, 21, 38, 54, 57, 68, 69
William Cocke, 165, 173
Cocks, 141
Bryant Cocker, Cooker, 103, 138
Benjamin Cockrum, 72
Henry Cockerham, 113
John Cockerham, 112
Phillip Cockerham/Cockrum, 43, 52, 69
Mrs Cockerham, 100, 106
James Cohoon, 76
Edward Colbreath, 96, 130, 135, 170, 173, 174
John Colbreth, 57
Peter Colbreath, 130, 135, 170, 174
Colbreath, 144[(2)]
John Cole, Coles, 31, 82, 112
Walter Coles, 169
William Cole, 49
Christopher Coleman, 110
James Coleman, 49, 64, 79, 86[(2)], 110, 151, 153
John Coleman, 110
Richard Coleman, 107, 121[(2)], 134
Robert Coleman, 110, 151
Howell Colleir, 164
Stephen Collins, 20, 23, 59, 85, 107, 108
Samuel Comer, 85, 97, 107[(2)], 109, 132, 171
Thomas Comer, 121, 122
Robert Conel/Connell, 91, 111
John Cooke, 59

William Cook, 51
John Cooper, 61
Samuel Cothman, 141
Clack Courtney, 89, 99, 164
Thomas Covington, 95, 97, 107, 128
William Covington, 59
Henry Cox, 107, 108
John Cox, 18, 33, 46, 49, 51, 56, 58, 61, 62, 65, 66, 67, 79, 105, 132, 134, 146, 151, 153, 157, 163
John Cox Junior, 162, 165
Peter Cox, 72
Capt. Cox, 56[(2)]
Cox, 138, 145, 172
Cozens, 143
Cornelius Cranshaw/Crenshaw, 47, 71, 112
Gideon Chranshaw, 59
Henry Crenshaw, 143
Joseph Cranshaw, 59
Thomas Crenshaw, 27, 71, 112
Hugh Creighton, 74
Edward Crews, 170
William Cross, 25, 38
Batt Crowder, 148
Stephen Crump, 106
Andrew Cunningham, 94, 95
James Cunningham, 67, 93, 110, 111, 163
John Cunningham, 67
William Cunningham, 94, 110, 155
George Currie, Curry, 18, 59, 88
Abraham Cuttilo, 76
Edward Cuttilo, 76
John Cuttilo, 76
Nanny Dabbs, 20
Richard Dabbs, 162
Timothy Dalton, 6
Dalton, 24
Chesley Daniel, 119, 169
Josiah Daniel, 169
Leonard Daniel, 119
James Daugherty, 111
Thomas Daugharty/Daugherty, 32, 110
James Dause, 126
Adlard David, 16
Samuel David, 78, 104
Baxter Davis, 153
Edward Davis, 39, 153

John Davis, 16, 64
Jonathan Davis, 47, 127
Joseph Davis, 52, 69, 82, 110
Robert Davis, 103, 161
Robert Davis's Son, 103
William Davis, 63, 99, 147, 148, 153
James Dawes, 27
Henry Decker, 86, 88
Scherer/Tscharner Degrafenreidt/Degrantenreidt, 37, 39, 50, 102
Henry Delony, 63, 64, 65, 66, 104, 111, 112, 117, 119, 145, 150, 158
Leeways/Lewis Delony, 5, $7^{(2)}$, 9, $12^{(2)}$, 17, 18, $19^{(2)}$, 28
John Denny, 29
James Denton, 129
John Dickerson, 94
William Dickerson, 110
James Dicks, 155
William Dobbyns, 27
Dockery, 31
David Dodd, 31
William Donathan, 126
Thomas Dorrity, 25
David Dortch, 73, 164
William Douglass, 16, 29, 98
William Dowsing, 168
Thomas Draper, 110
William Drew, 73, 96, 138, 147
John Dudgeon, 111, 168
Richard Dudgeon, 25, 69, 76, 111, 114, 132, 167, 168, 169
William Dudgeon, $111^{(2)}$, 115, 116
James Dukes, 129
John Taylor Duke, 110
Dun, 148
Dunnahoe, 8
Thomas Dupry, 81
Samuel Duval, 23, 72
Robert Dyer, 130
Robert Henry Dyer, 10
Edward Ealam – see Edward Elam
David Ealbanks - see David Eubank
John Earl, 143
Thomas Easly, 96
William Easley, 170, 174
John East, 111, 115, 116
Tarlton East, 143, 150
Thomas East, 111
William East Senr., 94

James Easter, 5, 40, 47, 68
Thomas Eastland, 18, 19[(2)], 34, 105
William Eastland, 110
William Eaton, 39
John Eckols, 55
John Edloe, 27
James Edmonds, 91
Thomas Edwards, 150
William Edwards, 50, 77
Edward Elam, 154, 162, 164, 172
William Elam, 154
Francis Ellidge, 28, 32, 42
George Elliot, 69, 81, 112[(2)], 117, 118, 128
Richard Elliott, 170
Edmond Ellis, 123
Jeremiah Ellis, 16
Richard Ellis, 103, 123, 156, 161
William Ellis, 103, 123, 161
Arthanatious Elmore, 129
Henry Embry, 24, 31, 35, 72
William Embry, 69, 80, 106, 107, 118
Edward Epps, 98
Thomas Erskine, 158
George Estis, 127, 140, 144
David Ealbanks (probably Eubank), 82
Daniel Hughbank (probably Eubank), 108
James Euin, 63
Stephen Evans, 65, 88, 91, 93, 101, 102, 144, 165
George Ezell, 91
John Ezell, 90, 91, 111[(2)]
Michael Ezell, 91, 111
Benjamin Farmer, 84, 96
Feild Farrar, 140, 151, 165
George Farrar, 147
Thomas Farrar, 110
George Farron, 53
Joel Ferguson, 115
James Ferrell, 158
Theophilus Field, 92
Martin Fifer, 49, 59, 70, 75
John Fillups – see under Philips
Daniel Firth, 21, 31
Isham Fisher, 165
Mrs Fisher, 37
John Fletcher, 59
John Flin, 60, 105

```
Loflin Flin  1
    i   Flin  1
    o     Fl nn 1
        Fon  in
Fon  in 8
 o n Fo
    o    Fo       1 4  8   1 8 1    1    148 14  1    1
    o    Fo        nio  1 8  148 14
          Fo    1   14  1   1   1   1     (2)
```
John Foster, 96, 108, 110, 112, 135, 136, 152
Thomas Foster, 97, 107, 123, 125, 130
William Foster, 139, 152[2], 154
Alexander Fowler, 156, 165
Richard Fox/Foxx, 8, 16, 64, 80, 98[2], 99, 151, 153, 156
John Francis, 46
Micajah Francis, 123
Nehemiah Frank, 38, 73, 94, 100, 104, 161
James Franklin, 111, 129
Peter Franklin, 94, 110, 143
Joseph Freeman, 152, 155
Luke Frolia (?), 126
William Fulsher, 62
John Fulton/Foulton, 111, 124, 156
John Fuquay, 74, 93, 110, 134, 171
Joseph Fuquay, 110, 143, 150
William Fuquay, 49, 110
William Gamblin, 98
David Garland, 61, 70, 73, 83, 105, 138, 158, 161, 162, 163, 165, 166[3], 167, 172
James Garland, 72
Nathaniel Garland, 72
Humphrey Garrett, 129
James Garrett, 129
Robert Garrett, 129
William Garrot, 30
Henry Gee, 74
Neel Gee, 74
William Gee, 50, 52, 60, 74, 155
David Gentry, 92, 93
Joseph Gentry, 100, 106
Nicholas Gentry, 100, 106
Simon Gentry, 92, 93, 100, 106
David George, 94, 95, 140, 141, 146[2], 155
James Gideon, 141
Henry Gill, 73, 105
Joseph Gill, 61
Michael Gill, 71, 86, 92[2], 152, 154

William Gill, 13, 92[(2)]
John Gilliam, 9, 66
William Gillum, 103
William Glading, 104
Mrs. Gladses, 111
John Glascock, 47, 71, 95, 97
John Glass, 86, 96, 105, 159
Jeremiah Glaunch, 49
John Goin, 104, 110
John Goin Junr., 104
William Goin, 104
Edward Goode, 61, 69, 80, 112
Jno ., John Good, 117, 141
Matthew Goode, 162
Mackerness/Mackness/McKness/McNess Goode, 12, 14, 36, 37, 44, 51, 54[(2)], 75, 78, 88, 91, 135, 162
William Goode, 103, 107, 126, 132, 142, 145, 151, 162, 165, 172
James Gordon, 76
Christopher Gormer, 51
Daniel Gorrie, 163
John Gorrie, 66, 67
Francis Graham, 90
John Grainger/Granger, 74, 99
Joseph Granger, 57, 99
Grayor, 132
John Green or Greer, 120, 129
John Greer, 111
Joseph Greer, 42, 43, 57, 61, 158
William Griffin, 155
John Grissel, 31
David Gwin/Gwyn, 20 21, 24, 41, 46, 114, 127, 129, 149
John Gwinn, 22, 39
Mary Gwin, 62, 74
Widow Gwin, 46, 58
Edward Haily, 112
James Haily, 112
John Haily, 112
Robert Haislop, 27
John Hall, 13
Moses Hall, 86, 108
Thomas Hall, 86
William Hall, 111
David Halliburton, 138
Thomas Hamlin, 123
John Handcock, 62
Thomas Handcock, 21, 75

Daniel Handkins, Hankins, 107, 108, 137, 152, 154
John Hankins, 69
William Hankins, 10
George Hannah, 74, 104
John Haney/Hany, 108, 152, 162
Gabriel Hardin, 31, 98
William Hardin, 31
William Hardwick, 110
William Hardwitch, 94
Charles Harris, 20, 51
Edward Harris, 103, 123, 161
George Harris, 87
Maynard Harris, 161
Samuel Harris, 50
William Harris, 5, 51, 60, 68, 97, 105, 120
William Harris (Finney Wood), 127
Benjamin Harrison, 16, 29, 98
John Harvie/Harvey/Harvy, 39, 103, 141
Thomas Harvie/Harvey/Harvy, 63, 78, 95, 139, 141, 148
William Harvey, 141, 147
Francis Harwood, 132
Colonel Harwood, 158
Stephen Hatchel, 104
Thomas Hatchell, 137, 138, 139
Jeremiah Hatcher, 26, 43, 60, 69, 80, 112
Robert Hatcher, 61
John Hatchet, 152
William Hatchet, 152
John Hawkins, 43, 80, 85, 108, 112$^{(2)}$, 145
Joshua Hawkins, 100, 106
Michell Hawkins, 112
Pinkethman Hawkins, 80, 82, 98, 106, 110
Richard Hawkins, 43
Thomas Hawkins, 48, 59$^{(2)}$, 102, 104, 106
William Hawkins, 17, 107
Hawkins, 129
Nicholas Hayle, 14, 15
Daniel Hayes/Haynes, 43, 53, 55, 68, 72, 122, 124, 134, 149
Henry Hayes/Hay/Hayse/Hays, 170, 171
John Hayse/Hays, 126, 152, 156, 157, 158
Matthew Hay, 158
Hays, 74
John Hazlewood, 143
John Hearndon, 131, 138
Robert Hester, 126
Humphrey Hewey – see Hughey, 31, 90, 98

Joseph Hickman, 9
Hickman, 36
James Hicks, 98
Joseph Hicks, 98
Richard Hill, 17, 20, 82
Thomas Hill, 129
William Hill, 27, 49, 102, 104, 124, 132, 148, 154
William Berry Hill, 67
Hill, 123
John Hight/Hite, 88, 131, 141, 155, 161[2]
Amos Hix, 123
John Hix, 100, 106, 144
Richard Hix, 24, 85[2], 86, 123, 128
Richard Hix Junr, 154
Thomas Hix, 140, 156
John Hobson, 43, 92, 93, 112, 145, 158
Nicholas Hobsom/Hobson, 43, 69, 112
Welcom William Hodges, 30
William Hogan, 10
Peter Holland, 55
Bennet Holloway, 123
George Holloway, 50, 63, 78[2], 79, 84[2], 123
Isaac Holmes, 104
Josiah Holmes, 110
Samuel Holms/Holmes/Homes, 46, 79, 149
John Holt, 153, 162
William Holt, 65, 136
Robert Hood, 126
Obediah Hooper, 112, 170
David Hopkins, 143
Michael Hoprick, 49
Henry Howard, 22, 50
William Howard, 5, 34
John Howell, 66, 130
Philip Hudgins, 51
Christopher Hudson, 61
Isaac Hudson, 68, 83
Peter Hudson, 12, 40, 57, 169
Daniel Hughbank – see Daniel Eubank
Anthony Hughes, 70, 75, 80, 98
Reese Hughes, 122, 166
Humphrey Hughey/Hewey, 31, 90, 98
Daniel Humphris, 58, 74, 102
John Humphries, 7[2], 57, 61, 127, 134, 135, 158, 164, 167, 168, 174
Humphries, 164, 168
William Humprey, 31

Charles Hunt, 93, 105, 110, 138
James Hunt, 21, 30, 45, 62, 74, 80, 82, 87, 93, 94$^{(2)}$, 96, 107, 110, 114, 125, 126, 134, 142, 148, 150, 153
Joseph Hunt, 93, 102
Joseph Hunt junior, 123
Memucan/Nenucan Hunt, 48, 50, 91, 116
William Hunt, 66, 97, 108, 120, 153
Captain Hunt, 129
William Hunter, 157
John Ingrum, 47
Charles Irby, 33
Christopher Irvin, 6
James Irvine, 34
Willian Irvin, 112
Henry Isbell, 26, 71, 81, 95, 97, 103$^{(2)}$, 113, 114, 133, 135, 140, 168
Edward Jackson, 27
Henry Jackson, 110
Hezekiah Jackson, 160
John Jeffrys, 153
John Jefferis Junr., 153
Ffield/Feilding Jefferson, 12, 13, 14, 31, 32, 35, 36, 53$^{(2)}$, 57, 58, 93
George Jefferson, 139, 143, 148, 155
Peter Jefferson, 147
John Jennings, 70, 76, 83$^{(2)}$, 126, 138, 158, 162, 167
Capt. Jinings's, 166
Jennings, 67$^{(2)}$
William Jeter, 165
Xr [Christopher] Johnson, 45
Isaac Johnson, 89, 117, 133, 137, 154, 161, 162
James Johnson, 47
John Johnson, 118, 119, 169
Joseph Johnson, 44, 64, 89, 154, 162
Michael Johnson, 31, 43
Miles Johnson, 169
Samuel Johnson/Johnston, 23, 58, 74, 87, 100, 109, 124, 136, 139$^{(2)}$, 140
William Johnson, 23, 58, 74, 100, 129, 135, 154
David Jones, 25, 74
George Jones, 154, 160
Godfry Jones, 40, 47, 107, 108, 148
Harwood Jones, 132, 156
Jacob Jones, 111
John Jones, 95, 144
Joseph Jones, 22
Philip Jones, 5, 25, 85, 86, 101, 107, 152
Reps Jones, 130, 141, 147
Richard Jones, 22, 86

Robert Jones, 9, 15
Samuel Jones, 38
Stephen Jones, 98
Thomas Jones, 5, 59, 60, 62, 72, 79, 86
Tignal Jones, 132, 138
Vinenal/Vinkler Jones, 132, 156
William Jones, 9, 56, 61
Alexander/Ellick Joyce, 65, 87, 95, 118, 120, 136, 163
George Keeling, 153
Leonard Keelin, 110
Osborne Keeling, 58, 74, 104
Kemp, 37
David Kenedey, 104
James Kidd, 169
John Kiersey Senr., 94, 110
John Keirsey junr., 94, 110
Thomas Kersy/Keirsey, 94, 110
Thomas Kersey Senior, 111
the Revd. William Key, Kay, 94[(2)], 100, 108, 123
Henry King, 153
John King, 39, 61
William King, 39
Samuel Kirk, 133
John Kitchen, 165
John Knight/Night, 37, 48, 164
Garrard Ladd, 98
William Ladd Senr., 98
William Ladd junr., 98
Samuel Lafon, 110
Mathew Lafoone, 129
Tidance Laine, 29
Claiton Lambert, 98
James Lambert, 98
John Lambert, 98
George Landford, 98
John Landford, 98
Thomas Lanear/Lanier, 8, 9, 49, 102, 104, 124, 132, 134, 138, 139, 140
Robert Langley, 141
Dennis Lark, 46, 58, 73, 98[(2)], 164, 166, 167, 171
Robert Larke, 149, 164, 167
Hugh Lawson, 19, 21, 25, 32, 33, 48, 49, 58, 115
William Lawson, 40
William Lax, 52, 69, 112
John Lee, 58
John Leeper, 104
Thomas Leigon – see Thomas Ligan

Robert Leveret – see Robert Liveritt
Edward Lewis, 126, 154, 157, 160, 173
William Lidderdale/Litherdale/Lydderdale, 33, 99, 152, 156
Thomas Ligan/Leigon, 108, 149
John Linton, 158
James Little, 111
Robert Liveritt/Leveret, 38, 55, 141, 147
David Logan, 16, 70, 75, 110, 140, 146, 152
John Logan, 30, 70, 75, 114
Richard Long, 126
John Love, 63
George Lovel, 63
Thomas Low, 100, 106
James Lowman, 76
Thomas Lowry, 83, 86, 121
John Lucas, 10, 97, 154, 155
William Lucas, 164
Abraham Lunderman, 58
William Mackadieu/McCadoe, etc., 43, 80, 112
William McClehomny, 74
Alexander McConnell, 110
Andrew Mc.Connell, 53
McConnico, 149
James McCoy, 104
James Mcdade, 23
James McDaniel, 33
Michael McDaniel, 22
Tarrance McDaniel, 111
Timothy McDaniel, 63
James McGlaughlin, 58, 110
James McKenny, 76
Michael McKie, 106
Michael McKie Senr., 100
James McMurday, 111
Daniel McNeil, 64, 91
Michael McNeil, 91
John McNess, 37, 111
Ephraim Mabry, 98
Joshua Mabry, 98[(2)], 102, 111, 151
Michael Mackie, 84, 174
Nicholos Major, 49
Stephen Mallet, 42, 54, 62, 63, 88, 137
Stephen Mallet Junr., 104
Daniel Malone, 46, 52, 81, 113
Drury Malone/Melone, 17, 98
John Mangram, 155

Francis Mann, 111, 115, 116, 162
Page Mann, 111
Samuel Manning, 12, 54, 91
Mathew Marable, 42, 49, 51, 52, 54$^{(2)}$, 55, 58, 63, 67, 77, 79, 87, 88$^{(2)}$, 93$^{(2)}$, 94, 97, 101, 102, 113, 116$^{(2)}$, 117, 118$^{(2)}$, 119, 120, 124, 134, 144, 145, 151, 152, 156, 169, 171
William Marable, 35, 36, 51, 54, 67, 75, 93, 102, 165
Mr. Marable, 87, 113
John Marshal, 63
Abra/Abram/Abraham Martin, 7$^{(2)}$, 9, 12, 14$^{(2)}$, 25, 26, 36, 38, 41, 44, 48, 49, 54, 56, 58, 59, 62, 66, 69, 73, 76$^{(2)}$, 79, 103, 170, 172
Andrew Martin, 31, 32, 62
Robert Martin, 104
Mr. Martin, 39
Charles Mason, 146
Mason, 43
Wilson Mattox, 58, 74, 104
Abraham Maury, 170
Daniel May/Mayse, 75, 84, 115, 141, 142, 144
Henry May/Mayse, 21, 68, 78, 89, 104, 141
Joseph Mayse, 22, 29, 34, 74
Mattox Mayse, 93, 110
William Mayse, 93, 110
Nicholas Maynard, 150, 155
John Mead, 6, 14, 23, 24
Abraham Merriman, 133
Conrod Messer Smith, 165, 168
Joseph Michaux, 169
Ensworth Middleton, 74, 104
John Middleton, 70
John Miller, 23
John Mills, 65
William Mills, 65
Joseph Minor, 24, 37, 121
Daniel Mitchell, 67, 91, 92, 155
Isaac Mitchell, 118
Jacob Mitchell, 66, 173
James Mitchell, 32, 35, 36, 41, 45, 48, 111, 114, 171
John Mitchell, 63, 152, 154$^{(2)}$, 162, 164
Robert Mitchell, 67
Thomas Mitchell, 71, 86
William Mitchell, 98, 124, 168
Capt Mitchell, 27
John Mize, 91, 98, 111
Benjamin Mobberly, 47
Clement Mobberly, 39

Edward Mobberly, 39, 47, 55
John Mobberly, 47
Samuel Moon, 63
John Morgan, 143
William Morgan, 14
Drury More/Moore, 76, 172
George Moore, 27, 82, 110, 112, 137, 160
Hugh Moore, 15
Joseph Moore, 16
Thomas Moore, 106, 110[2], 138, 140, 151, 165[2]
Thomas Moore Jun., 165
More, 96
Philip Morgan, 31
Reuben Morgan, 31, 51, 54, 110
William Morgan, 23
Joseph Morton, 5, 70, 75, 85, 95, 109, 120, 124, 128, 131, 151
Joseph Moreton Junr., 118
Josiah Moreton, Morton, 118, 159
Samuel Moreton, 51, 137
Edward Mosby, 152, 154, 173
George Mosly, 50
David Moss, 173
William Moss, 47
John Mount, 20, 58, 74
John Mullins, 20
Valentine Mullins, 59, 126
Robert Munford, 126, 145, 162, 165
Barnaby Murphey, 16
James Murphy, 104
John Murphey, 155
William Murphey, 16
Jeffery Murrel, 72, 143
John Naull, Nall, 155, 156, 166
Richard Naul, 136, 142, 155, 156
Frederick Nance, 112, 159
John Nance, 6, 28, 47, 112
John Nance Senior, 71
John Nance Jur., 27
Richard Nance, 47, 71, 112
Richard Nance Junr., 27
Thomas Nance, 27, 47, 71, 123, 146, 149, 160, 168
William Nance, 47, 71, 112
William Nance junior, 71
Thomas Nash/Nast, 28, 40[2], 68, 74, 80, 84, 100, 108, 118
Thomas Neal, 162, 165
James Nelson, 87

Thomas Netherly, 158
Julius Nichols, 31, 36, 51, 61, 80, 82, 90, 98, 110
William Nichols, 31
William Niel, 66
John Night – see John Knight
Hugh Norwell, 110
Thomas Norwell, 110
John Oliver, 155
Matthew Organ, 131, 155
James Orr, 104
John Orr, 104
Reps Osborne, 162
Peter Overbey, 120
John Owen, 136
William Owen, 8
William Owl/Owle, 63, 141
Jessee Ozling, 150
John Pallet (Paulett, as properly pronounced), 139, 141
Richard Palmer, 59, 91[2], 120, 126, 140
Michael Parengame, 49
Charles Parish/Parrish, 130, 142
John Parish, 129
Peter Parish, 45
John Parker, 8, 129
James Parrott, 88, 124, 126
Christopher Parsons, 78, 95, 103
Richard Parsons, 22, 38
Jonathan Patteson, 163
George Pattitor/Pattillo, 114, 159, 168, 169
John Patrick, 110
John Payne, 47
Samuel Peace, 171
William Pearson, 129
William Pennil, 111
John George Pennington, 90, 91, 111
Joseph Perrin, 46, 47, 62, 66, 100, 108
Samuel Perrin, 26, 41, 46, 62, 87
William Perrin, 62
William Perry, 50, 170, 174
John Pettus, 131
Thomas Pettie/Pettis/Pettus, 69, 81, 112, 128, 132
Petis, 123
Francis Petty, 86, 137, 138, 152, 159, 164
Francis Moore Petty, 44, 71[2]
William Petty, 72
John Phelps, 19

Samuel Phelps, 120, 126
William Philby, 111
George Philips/Phillups, 73, 133
John Fillups, 165
Israel Pickens, 25, 31
William Pinnell/Pinner/Pinnin, 90, 91[(2)]
Thomas Pitman, 47
George Platt, 69
John Pleasants, 87, 100, 103
Philip Poindexter, 43, 52, 69, 112, 158
Benjamin Pollard, 143
Adam Poole, 167
William Pool/Poole, 81, 134
Peter Petty Poole, 112
William Petty Pool, 71, 114, 137
Thomas Portwood, 106, 107
John Potter, 138, 140, 154, 157
Thomas Pound, 143
Edward Powel, 46
John Pratt, 23, 85
Michael Prewit, 94, 105
Richard Prewit, 94, 105
Reese/Rice Presse/Price Senr., 24, 45, 47
Thomas Price, 47, 104
William Price, 163
Cornelius Priest, 135
Epharim Pucket, 143
Shippe Allen Puckett, 153
Spettle Pully, 84, 90
John Puryear, 151, 165
Benjamin Raden, 141
Baxter Ragsdale, 135
Benjamin Ragsdale, 68
Crooked Creek Ragsdale, 74
John Ragsdale, 68, 83, 86, 97, 105, 135, 150, 165, 167[(2)], 172
John Ragsdale Senior, 76
John Ragsdale Junior, 76
Joseph Ragsdale, 61
Richard Ragsdale, 158
Thomas Ragsdale, 154
Josias Randle, 47
Richard Randolph, 24
Colo. Randolph, 5
Randolph, 141
Peter Rawlins, 157, 162
Francis Ray, 27, 83, 126

Ruben Rey/Roy, 20, 38
Clement Read/Reed, 5, 13, 19, 21, 23, 47, 48, 52, 58, 60, 62, 68, 74, 76, 79, 85, 86, 100, 101, 103, 109[2], 114[2], 118, 121, 124, 132, 139, 160, 162
Clement Read Junr., 108
John Read, 20
Thomas Read, 154
William Read, 170
William Redman, 47, 75
Clement Reed – see Clement Read
Joseph Rentfro, 23
Thomas Rice, 20
William Richardson, 64, 108
John Richdal, 97
John Richey, 63
Robert Richey, 63
William Riddle, 80, 82, 90, 102, 112
William Rivers, 161
Rivers, 72
Jasper Robernett, 146
David Roberts, 86, 139, 152
James Roberts, 47, 52[2], 81[2]
James Roberts Senior, 71, 112
James Roberts Junior, 71, 112
John Roberts, 105
Thomas Roberts, 110
William Roberts, 50, 117
Alexander Roberson, 110
Henry Roberson/Robertson, 50, 165
Israel Robertson, 99
Jacob Roberson/Robinson, 87, 110, 123
John Roberson, 94, 165
Nathaniel Robertson, 66, 118, 119
William Roberson, 94, 117
Jacob Robinson Senr, 94
John Robinson, 128
William Robinson, 127, 143, 144
Richard Rockett, 128
Andrew Rodgers, 110
John Rodgers/Rogers, 110, 158
Thomas Rodgers/Rogers, 32, 110, 143
William Rodgers/Rogers, 20, 23, 25, 32, 67
Augustine Roling, 68
John Roling, 68
John Ross, 143
Henry Rottenbury, 16
William Roulet, 152

Peter Rowlet, 27
Jacob Royster, 105, 119, 144, 157, 159
William Royster, 64, 81, 82, 88, 108, 119, 130, 135, 157, 159, 170, 173
Joseph Rudd, 165
Samuel Rudd, 129
Samuel Rudder, 126
William Ruffe, 151
John Ruffin, 104
Capt John Ruffin, 20
Colo. Ruffin, 54
Jeffrey Russel, 33, 45
John Russell, 108
Philemon Russell, 15
William Russell, 30
James Rutherford, 90
Thomas Rutherford, 47
Isaac R_____, 143
William Saffold/Soffold, 27, 49, 61[2], 74, 92, 97
Henry Sage, 32, 49
James Sandifer, 173
John Sanford, 135
John Sansum, 38, 103
Richard Sansun, 62
John Satterwhite, 156
Michael Satterwhite, 96
Thomas Satterwhite, 59[2], 120
John Saunders, 34
George Scot, 104
John Scott, 137, 144, 165
Scott, 35, 128
Richard Scruggs, 91
John Segant, 98
Thomas Shelborne, 112
James Shelton, 152, 160
Ralph Shelton, 121, 131, 145, 156, 164
Richard Ship, 61, 62, 79, 118, 158
Wm. Shorte, 133
John Silcock, 103
John Simmons, 39
Arthur Slaton/Slayton, 111, 129, 146, 162
Daniel Slaton, 111
Wm. Slaughter, 133
Benjamin Smith, 123, 161
Guy Smith, 153
John Smith, 108, 111, 143[2], 150, 153, 158, 161[2]
Luke Smith, 16

Thomas Smith, 123, 172
Micajah Smithson, 123
Christopher Sneed, 81
Henry Soan, 27
William Soffold – see William Saffold
James Sparrow, 98
Austin Spears, 91, 92
John Speed, 46, 50, 60, 84, 89, 102[(2)], 111, 112, 117, 119, 139, 141, 143, 145, 150, 155, 158, 171, 173, 174
Thomas Spencer, 5, 118, 148, 149, 162, 170
Thomas Spraggons, 82, 93
William Spraggons, 82, 93
Marmaduke Stanfield, 58
Thomas Standfield, 111
William Stanfield/Standfield, 58, 74, 85, 111
Thomas Staples, 28, 47
William Steith, 67
John Stephens, 169
Thomas Stephens, 169
James Stewart, 42, 59
John Stewart, 20, 111, 121, 170
Thomas Stewart, 70, 75, 111
Jacob Stober, 20
Allen Stokes, 163
Charles Stoke, 143
David Stokes, 11, 13, 21, 25, 53, 56
Richard Stokes, 27, 52, 80
Silvanus Stokes, 27
Silvanus Stokes Sr., 64
Young Stokes, 6
Stokes, 46
William Stone, 45, 47, 69, 82, 112
Joel Stow, 104
John Strann, 143
John Strone, 72
Richard Sullins, 110
Charles Sullivant, 47, 81, 154, 162
Daniel Sullivant, 20
James Sullivant, 47, 106, 127
John Sullivant, 161
Menoah Sullivant, 106
Owen Sullivant, 20, 51, 62, 72, 85, 103
Richard Swepstone/Swepsons 152, 153, 155, 167, 174
Thomas Tabb, 117, 118, 119, 120, 126, 129, 132, 135, 144, 162, 166, 169, 172
Hezekiah Taber, 126
William Tabor, 15[(2)]

Charles Talbot, 16, 21, 22, 30, 39, 44, 47, 59, 62, 75
Mathew Talbot, 8, 18, 47
Abraham Talley, 45
Henry Talley/Tally, 27, 45, 104
John Talley, 45
Branch Tanner, 93, 102
Lewis Tanner, 99
Mathew Tanner, 59[(2)]
Thomas Tanner, 99
Thomas Tate, 59
Edmund Taylor, 119, 125, 150, 154, 156, 157[(2)]
James Taylor, 41, 72, 81, 100, 102, 106, 109, 124, 128, 132[(2)], 133, 139, 150, 151, 153, 159, 167, 168[(2)]
John Taylor, 98
Joseph Taylor, 129
Samuel Taylor, 158
Thomas Taylor, 108
Thomas Taylor Senr., 98
Thomas Taylor junr., 98
William Taylor, 163, 165, 166[(2)]
Andrew Tecker, 49
John Templin, 104
Nathaniel Terry, 47
William Thomas, 20, 29, 66
James Thomason, 133
William Thomason – see William Thompson
Thomason, 146
John Thompson/Tomson, 33, 45, 64, 86, 105, 134
Richard Thompson/Tompson/Tomson, 27, 45, 71
Wells Thompson, 173
William Thompson/Thomason, 67, 111, 123, 169, 172
John Thornton, 143
Mark Thornton, 81
Thomas Thornton, 112
Thornton, 168
Patrick Tilen, 47
James Timker, 35
Amos Timms, 31, 99
Manoah Tinsley, 154
Edmund Toombs, 136, 152, 154
Gabriel Toombs, 130, 135
William Toombs, 154, 162
Joel Towns, 46, 62, 103, 122
John Towns, 21
Thomas Trammell, 129
William Traylor, 45

William Traynum, 154
Richard Treadway, 58
John Trice, 106
James Tucker, 61
Warner Tucker, 61
William Tucker, 91
John Turner, 47
Mathew Turner/Turnes, 140, 157, 160
John Twitty, 13, 26, 31, 43, 52, 69, 70, 75
Thomas Twitty, 9
William Twitty, 80
Twitty, 67
John Upcott, 63
John Ussery, 63, 141
John Vance, 64, 111
Abraham Vaughn, 103
George Vaughn, 64, 73, 133[(2)]
Thomas Vaughn, 146
Henry Vandycke/Vandyke/Vandyck/Venduke, 126, 163, 165, 166, 170, 174
William Verdeman, 7, 35
Isaac Vernon, 111
James Vernon, 111
Jonathan Vernon, 84, 111
Thomas Vernon, 23, 78, 101
Thomas Vernon junior, 72
Jeremiah Viditto, 11
Andrew Wade, 33
Hampton Wade, 18, 50, 52, 56, 79, 102, 149
Robert Wade, 34, 67
Robert Wade Sr, 50, 58
Bozwell Wagstaff, 173
Francis Wagstaff, 92, 113, 125
Michael Waldrope, 95
James Waldun, 62
Anothony Walke, 52
Silvanus Walker, 43, 64, 82, 116, 119
Tandy Walker, 33, 50
William Wallace/Wallice/Wallis, 76, 83, 86, 90, 105
William Wallace's son, 76
Edward Waller, 142, 171[(2)]
John Waller, 87, 104, 114
Elisha Walling, 22
George Walton, 27, 39, 50, 71, 92[(2)], 98, 102, 112, 152, 159, 164
Sherwood Walton, 146, 160
John Ward, 111, 167, 168
Richard Ward, 23, 87

Seth Ward, 111, 152, 167, 168
Wade Ward, 98
Thomas Watkins, 94, 105
William Watkins, 82, 100, 126, 160
Duglass Watson, 147, 148
John Watson, 17, 104
Mathew Watson, 139, 141, 148, 162, 168, 171
William Watson, 87
Robert Weakley/Weakly, 129, 142, 146, 153
Charles Weatherford, 45
Jno./John Weatherford, 45, 63
Wm. Weatherford, 45
John Philip Weaver, 55
Merry Webb, 38
Michael Weeks, 31
Thomas Weeks, 31
Barnabas/Barnaby Wells, 47, 71, 112
Elijah Wells, 71, 112, 159, 160
George Wells, 44, 71, 96, 102, 108, 112, 137, 138
John Wells, 112
Thomas Wells, 54
Joshua Wharton, 103, 161
Edward Whit/Whitt, 27, 33
Elisha White, 139, 141, 150, 151, 153, 162
John White, 139, 143$^{(2)}$, 162
William White, 57, 61, 157
Benjamin Whitehead, 158
Richard Whitten/Witton/Wilton, 31, 32, 33, 48, 49$^{(2)}$, 53, 58$^{(3)}$, 60, 69$^{(2)}$, 79, 83, 89, 108, 116,
    126, 129, 132$^{(2)}$, 134$^{(2)}$, 152, 158
Colo. Witton, 96, 100, 158
Abraham Whittemore, 166
John Wilborne, 33, 68, 97, 105, 108
Benjamin Wilks, 112
James Wilkins, 59, 126
John Wilkins, 59
Richard Wilkins, 126
Thomas Wilkins, 59
George Williams, 50
Henry Williams, 117$^{(2)}$, 137, 140
James Williams, 133$^{(2)}$, 161
John Williams, 18, 19, 42, 70, 107$^{(2)}$
Joseph Williams, 27, 31, 89, 102, 110, 116, 119$^{(2)}$, 122, 124, 132, 134, 161
Lazarus Williams, 10, 17, 72, 122
Mathew Williams, 66, 129, 135
Peter Williams, 153

Richard Williams, 72, 142
Robert Williams, 21, 38
William Williams, 28$^{(2)}$, 30, 37, 50, 51, 54, 76
Thomas Williamson, 66, 68, 81
Jarrel Willingham, 142
John Willingham/Winningham, 40, 47, 71
John Willingham Junr., 27
Thomas Willingham/Winningham, 45, 47, 71$^{(2)}$
William Willingham, 71
Wm, Willingham junior, 71
Winningham, 45, 46, 47
Edward Willis, 105
Steven Willis, 68
Thomas Willis, 63
George Wills, 110
Wilmoth, 145
John Wilson, 31
Peter Wilson, 10
Robert Wilson, 100, 106, 112, 113
Samuel Wilson, 15, 19, 130, 135
William Wilson, 42, 52, 100, 106, 112
Samuel Wimbush, 160
Adam Winder, 61
Thomas Wise, 33
Richard Womack, 118
Robert Womack, 5, 21, 40, 78, 86, 99, 100, 118
John Wood, 63, 111, 141
Robert Woods, 37, 58, 74$^{(2)}$, 78, 104, 109
Stephen Wood, 89, 154, 161, 162
Francis Worsham, 86
John Worsham, 154
Robert Worsham, 152, 154
John Worthy, 75
Thomas Worthy, 21, 31
John Wright, 151, 173
Thomas Wright, 173
Richard Wyatt, 129
Hugh Wyley, 76
Daniel Wynne/Wynn, 115, 134, 143
John Wynn, 149
Robert Wynn, 41
Thomas Wynne, 18, 58, 139, 144, 149
William Wynne, 11
Charles Yancey, 137
James Yancey, 64
Richard Yancey, 135, 170, 174

John Young, 12, 14, 18, 20
Peter Young, 110
Samuel Young, 45, 160, 173

Bridges

Bridge, 47, 87, 116[(2)], 163, 171
Allen's Creek Bridge, 12, 35, 46, 63, 117, 122, 138, 140, 171, 172[(2)]
Banister Bridge, 41, 43, 46
Banister River bridge at Cowford, 25
Bleu/Blew/blue Stone Creek bridges, 36, 48, 50, 101, 126, 128, 130, 153
Bollings Bridge, 159
Cocks Creek Bridge, 62, 63, 76
John Cox's Bridge, 58, 79, 131, 134, 146, 163[(2)]
Crooked Creek Bridge, 33
William Crosses Bridge, 38
Cubb Creek bridges, 21, 27, 37, 48, 69, 76[(2)], 79, 80, 87, 94, 96, 107, 126, 131, 132, 134, 151, 157, 159, 163, 168
Little Bridge below Cub Creek, 152
flat rocke Creek bridges, 138, 162, 166, 170, 174
F****ing Creek bridge, 21, 163
great Creek Bridge, 36, 121
Gwinns Bridge (Little Ronoak), 21, 122
Hatchers Bridge (north Meherrin), 126
Hurts Bridge (Hust Bride?), 141, 147
ledbetter Creek bridge, 25, 43
Louse Creek Bridge, 45, 58, 125, 132, 140, 141, 150
Mathew Marable's Bridge, 97, 101
Maherrin/Meherrin River bridges, 17, 18, 19[(4)], 27, 30, 31, 35, 41, 43, 45, 46, 47, 48, 49[(2)], 50[(2)], 51, 53, 56, 58, 61, 62, 80, 82, 84, 116, 119, 122, 126, 127, 128, 129, 132, 135, 140, 146, 157, 163, 169[(2)]
Miles's Creek Bridge, 9, 12[(2)], 60, 89, 91, 119
Mises/Mize's ford Bridge, 58, 150
Mortons/Morteons Bridge (Little Roanoke), 128, 151, 159
Mossing Foard bridge, 68, 76, 83, 86, 122, 125, 129, 144
North River Bridge, 19
Nottoway River bridges, 11, 25, 33, 43, 44, 48, 50, 79, 118, 120, 121, 128, 145, 149, 156[(2)], 160, 164
Rays Bridge, 27, 28
Reedy Creek Bridge, 17[(2)]
Roanoke Bridge, 145, 156
Little Ronoke Bridges, 17[(2)], 18, 20, 21, 23, 26[(2)], 28, 51, 52, 58[(3)], 65, 74, 76, 79[(2)], 84[(2)], 85, 95[(2)], 97, 100, 102, 109, 114[(2)], 115, 118[(2)], 120, 121, 124, 128, 131, 132, 133, 151, 159
Roberson Creek Bridge, 46

Scotts Bridge (Maherin), 172
Stokes's Bridge, 38, 130, 133
Stokes's Mill Creek Bridge, 94
Swish Creek Bridge, 127[(2)]
Turnip Creek Bridge, 30, 43, 114[(2)], 115, 143, 150
bridge near Hampton Vades, 52, 79
Silvanus Walkers Bridge at Scotts foard on northfork Maherrin, 64, 119
Wards fork bridge, 21, 32, 34, 58, 60, 78, 90, 125, 132[(2)], 150, 163, 171
Willinghams Bridge, 71, 91, 98, 102, 109, 164

Chapels, Churches, Glebes, Parishes

Allens Creek Church (?), 64
Ash Camp Church, 152, 159, 160, 164
Bouldings Church, 125
Burgamy's Church, 172
Chapel, Church, 7, 9, 21, 40, 55, 72, 88, 90, 100, 101, 133, 165
Cornwall Parish, 113[(2)], 153, 162, 170, 171, 172[(2)]
Cornwall Parish Church, 170
Court House Church, 131
Cub Creek Church, 150
Cumberland Parish, 113, 150, 172[(2)]
Cumberland Parish Glebe, 150, 158
Cumberland Parish line, 152, 159, 160, 164
flat Rock Church, 126, 130, 135, 141, 142, 147, 150, 158
Jefferson Church, 110
the new Church, 143, 149, 162
Church on Otter River, 39, 55
Ready/Reedy Creek Church, 56[(2)], 70, 71, 92, 93, 100, 117, 121, 122, 134, 135, 142, 143[(2)], 150, 158
Ronoke Church, 60, 171
Little Roanoak Church, 92[(2)], 121, 131, 145, 172
Saint James Parish, 166, 167, 172[(3)]
Sandy Creek Church, 170

Ferries

William Abbots Ferry (Ronoke), 53
Thomas Anderson's Ferry (at Mitchell's Ferry on Roanoak), 91
Richard Blanks's Ferry (Berry's Ford on Roanoke), 118
Blank's Ferry, 117, 125, 136, 142, 169
James Blanton's ferry (Roan Oak opposite to Fox's Landing), 169
Bookers ferry, 134, 143
John Boyd's Ferry (Dann), 24, 34, 40

Robert Brookhouse's ferry petition, 30
John Camps Ferry, 144, 161, 173
Cargills Ferry on Staunton River, 5, 8
George Currie's Ferry over Great Ronoke from Munfords Quarter to the Ochenechee, 18
Adlard David's Ferry, 16
Ferry Rates, 9, 13, 16, 18, 24[(2)], 29, 34, 42, 45, 49, 66, 69, 74, 80, 81, 88
Peter Fontaine's ferry (Stanton), 69, 89
Richard Fox's Ferry/Fox's landing (Ronoke), 64, 98[(2)], 99, 151, 153, 156, 169
John Fuquay's Ferry (Stanton), 74
William Fuquay's Ferry (Stanton), 49
Fuquays Ferry, 140, 141, 143, 150[(2)], 152, 167, 168
Hampton's Landing, 13
Colonel Harwoods ferry, 158
Hawkins's fferry, 29
Hickmans Ferry, 9, 36
James Hunt's Ferry (Stanton), 80, 82, 93, 94[(2)], 105
Field Jefferson's Ferry (Ronoke), 13, 14, 50, 53[(2)], 72, 76, 78[(2)], 90, 93
Kemp's Ferry, 154
Joseph Mayse's late Ferry landing, 74
Mayses's Ferry, 93
James Mitchells Ferry (to Thomas Anderson's), 45
Capt. Michells Ferry, 106
Mitchells Ferry, 8, 83, 86, 91, 96
Capt Mitchells Landing, 27
Julius Nicholss Ferry (Stanton River, formerly Hickman's), 31, 36[(2)], 39, 80, 82, 98
Peter Overbey's (John Green's) Ferry, 120
Palmers Ferry, 120
Palmers landing on Butchers Creek, 59
Richard Randolph's Ferry (Staunton, "Roanoak"), 24
William Royster's Ferry (Ronoke), 81, 88, 119, 130[(2)], 131, 135, 144[(2)], 170, 173, 174
William Roysters Ferry Bond, 82
James Steward's/Stewart's Ferry (Stanton), 42, 47, 59, 70, 75, 84, 121, 136
Taylors Ferry (Roanoke), 159, 168, 172[(2)]
William Thomas's Ferry on Stanton River a mile below Long Island, 66, 67
Thomas Twitty's Ferry (Horse Ford on Roanoke), 9
Twittys Ferry, 16[(2)]
Robert Wade's Ferry (Stanton at Cornelius Cargill's), 34, 36, 41, 43, 46, 127, 136, 149, 154
Robert Wade's Ferry Boat, 58
Francis Wagstaffs Ferry Landing (Roanoke), 113

Fords

Akins's ford (Butchers Creek), 106, 151
Berry's Ford (Roanoke), 118

Cargill's Horse Ford (North River), 10, 11[(2)]
Cow Ford (Banister), 25
Cub Creek Old ford, 169
Cunninghams fford, 13
Davis's ford, 86
Dudgeon's ford, 126
flat rocke old ford, 138
Fish Dam Ford (Otter River), 6
Fuqua Ford, 30
Graves's fford, 12, 40
Grymes's foard, 129, 142, 151
Horse Ford (Banister), 16
Horse Ford (Ronoke), 8, 9
Peter Hudsons foard (Stanton), 57[(2)], 78
Christopher Irvins Ford (Otter), 6
Kings Ford (Ronoke), 8, 9, 49, 88, 118
Martins ford, 18
Mays's fford (Stanton), 13, 30
Robert Mitchells Ford, 16
Mitchells Ford, 9, 43, 91, 92
Mizes/Mices/Mises Ford (Maherrin), 17[(2)], 30, 46, 53, 58, 63, 64, 90, 91, 104, 139, 150, 151
Mossing ford, 16, 21, 39, 44, 47[(2)], 59, 62, 68, 76, 84, 86, 108, 113, 122, 125, 129, 144
Ford on Otter River Near the Mouth of Elk Creek, 35
Palmers ford, 57, 126
Parrishes fford, 125
ford over Meherrin at Francis Rays, 83
Rutledges ford (Appomattox), 17, 20
Saffolds ford (Meherrin), 83, 135, 148, 163, 166, 170, 171, 174
Scotts fford (Maherrin), 10, 12, 64
Shorts fford (Goose Creek), 14
Silvanus Walker's Ford (Meherrin), 119
Winninghams Ford (Maherrin), 6[(2)]

Houses

John Bacon's house, 74
William Been's House, 29
David Caldwell's House, 106
John Cox's House, 163
George Elliott's House, 117
Pinkethman Hawkins's House, 80
Henry Isbell's House, 81
Isaac Johnson's House, 154
Mathew Marrable's house, 52, 54, 88, 101, 124, 156

Abraham Martins House, 103
Abraham Maury's House, 170
John Owen's House, 136
Clement Read's House, 100
George Waltons dwelling House, 92[(2)]
Joseph Williams's House, 124
Wimbish's house, 154
Richard Wittons House, 134

Mills

Colo Birds Mill, 97, 105
Thomas Bedford's Mill, 161
William Calloway's Mill, 6, 35
Cox Mill, 74
Delonys Mill, 28[(2)], 62
Freemans Mill, 49
Lawsons Mill, 40
Marables Mill, 122[(2)]
Joseph Morton's Mill (Little Ronoke), 7
John Phelp's Mill, 19
James Roberts's Mill, 81[(2)]
Israel Robertsons Mill, 99
Cola. Ruffins Mill, 54
Ruffins old mill, 84
Stokes's Mill, 94
Trumans Mill, 135
John Young's Mill, 12, 14
Youngs Mill, 9, 14, 16, 17, 18, 21, 23, 24

Mountains and Hills

Blew ridge, 65
Johnson's Mountain, 35
long Mountains, 6
Stony Hill, 96, 103, 153, 154
Wart Mountains, 22[(4)]

Ordinaries

Clarks Ordinary, 159
Claunch's Ordinary, 86[(2)], 88
Coxes Ordinary, 172

Deloneys old Ordinary, 42, 43, 46, 55, 64, 88
Fosters Ordinary, 142
Gills ordinary, 134
John Humphris's Ordinary, 147, 174
Capt. John Jennings's Ordinary, 83
Jennings's Ordinary, 61
William Marable's Ordinary, 36[(2)]
Marrables Ordinary, 63, 75[(2)]
Twitty's Ordinary, 32, 33
Wilmoths Ordinary, 159, 165, 172

Plantations

Baxter's Plantation, 39
Hugh Boston's Plantation, 70
Sherwood Bugg's Plantation, 171
William Caldwell's Plantation, 115[(2)], 116
David Christopher's Plantation, 147
Abraham Cook's Plantation, 25
Edward Davis's Plantation, 39
Richard Dudgeon's Plantation, 167, 168, 169
Dudgeons Plantation, 152
Edward Ealam's Plantation, 162
Richard Elliott's Plantation, 170
Peter Fontaines Plantation, 89
Daniel Haynes Place, 43
John Humphries's Plantation, 158
Samuel Jones Plantation, 38
William Jones's Plantation, 9
William Keys (Kays) Plantation, 94, 100
the Revd Mr. William Keys Plantation, 94, 108, 123
David Logans Plantation, 146
Thomas Lowry's Plantation, 83, 86
Mathew Marables Plantations, 93, 94, 97, 113, 116
Daniel Mitchells Plantation, 155
William Mitchells Plantation, 98
Robert Munfords plantation, 145, 162, 165
Jonathan Patteson's Plantation, 163
the Pocket, 35
Clement Read's Plantation, 100, 101, 124, 1332
Guy Smith's Plantation, 153
John Speed's Plantation, 171
James Taylor's Plantation, 132
Wells Thompson's Plantation, 173
George Walton's Mill Plantation, 71

Whites Plantation, 106
John Williams's Plantation, 107
Richard Witton's Plantation, 89

Quarters

Roger Atkinson's Quarter, 166
Stephen Bedfords Quarter, 47
William Bookers Quarter (Mason's Creek and on the Little Beaver on Pond), 52, 60, 61
Augustine Claiborne's Quarter, 166
Cockes Quarter, 74, 140
Johnsons Quarter, 27
Masons Quarter, 43
Millers Quarter, 51
Joseph Mortons Quarter (Staunton), 70, 75
Munfords Quarter, 18
Thomas Nash's (Spring Creek), 108
John Paynes Quarter, 47
John Pleasants Quarter, 103
Clement Read's Lower Quarter, 62
Clement Reads uper Quarter, 86
Capt John Ruffin's Quarter, 20
Ruffins Quarter, 131
Smiths Quarter, 72
Tusling Quarter, 70, 75

Rivers, Creeks, etc.

Aarons Creek, 16, 57, 120, 144
Allen's Creek, 5[(2)], 12, 22[(2)], 26, 28, 34, 35, 40, 63, 64, 78, 90, 105, 106, 117, 122, 132, 138, 140, 151, 162, 164, 168, 171, 172[(2)]
Appomattox River, 6, 20
Ash Camp Creek, 5, 11, 83[(2)], 96, 97, 152
Banister River, 10, 11[(2)], 16, 22[(2)], 25, 29, 30, 43
lower falls, Banister River, 29, 30
Bear Creek, 147, 148
Little Bair Creek, 77, 78
Bairs/Bares/Barrs/Bears Element Creek, 26, 42[(2)], 61, 70, 90, 172
Bear Skin, 38[(2)]
Little Beaver Pond, 33, 61
Black Water, 14, 15
Blue/Blew Stone Creek, 5[(2)], 15, 32, 33, 36, 48, 50, 60, 66, 101[(2)], 102, 117[(2)], 126, 128, 130, 153
Great Blue Stone, 117

Breedloves Creek, Breedloves fork of the Juniper, $44^{(2)}$, 110, 112
Briery/Bryery River, 72, 78, 82, 118, 131, 160
Buckhorn, 8, 64, 65
Buffelo Lick, 9
Butcher's Creek, $5^{(2)}$, 34, $59^{(2)}$, 105, 106, 126, 138, 154, 165
little Cheary Stone/Cherry Stone, 26, 38
Cocks Creek, 62, 63, 76, 88, $104^{(2)}$
Cox Creek, 54
County line Creek, 33
Crooked Creek, 33, 44, 61, 72, 163
Cubb Creek, 7, 13, 19, 20, 21, 27, 37, 48, 69, $76^{(2)}$, 78, 79, 80, $87^{(2)}$, 90, 94, 96, 107, 115, 126, 131, 132, 134, $146^{(2)}$, 150, 151, 152, 157, 159, 167, 168, 169
Dann River, 10, 24, 33
Dockerys Creek, 164
Dorches Creek, 110
Double Creeks, 10, 11, 15
Dry Creek, 143
Dunavant Creek, 84
Elk Creek, 35
Elk horn Creek, 30
Falling River, 5, 8, 19, $20^{(3)}$, 23, 25, 29, 63, 77, 95
Finnywood/Finny Wood Creek, 27, 44, 127, 162, 165
fishing Creek, 33, 34
flat Creek, 54, 164
flat Rock Creek, 42, 49, 121, 129, 130, $135^{(3)}$, 138, 162, $166^{(2)}$, 170, 174
F****ing Creek, 18, 163
Goose Creek, 14, 35, 39, $55^{(2)}$
Grassey/Grasty Creek, 130, 135
Great Creek, 36, 121
Haw Branch, 133
Horse Pen/Horspen Creek, 59, 60, 66, 72, 94, 100, 169
Horsepen branch (fork) of Juniper Creek, $71^{(2)}$
Hound's Creek, 38
Island Creek, 155, 156
Junaper/Juniper Creek, 64, $71^{(2)}$, 131
Kettlestick Creek 49
Kittle Creek, 153
Leatherwood Creek, 37, 38
Ledbetter Creek, 25, 43, 71, 143
long Island, 66
Louse Creek, 45, 58, 125, 132, 140, 141, 150
Maggotty Creek, $23^{(2)}$
Maherrin/Meherrin River, $6^{(2)}$, $10^{(2)}$, 12, $15^{(2)}$, 17, 18, $19^{(4)}$, $26^{(2)}$, $27^{(2)}$, 31, $32^{(2)}$, 33, 35, 36, $41^{(3)}$, $42^{(2)}$, $43^{(2)}$, 45, 47, $48^{(2)}$, $49^{(2)}$, $51^{(2)}$, 52, 53, 56, 58, 61, 62, 63, $64^{(2)}$, 71, 80, 82, $83^{(3)}$, 84, 91, 96, $98^{(3)}$, $102^{(2)}$, 116, 117, $119^{(2)}$, 121, 122, 124, 127, 129, 130, 132, $135^{(2)}$, 138, 140, 142, 146, 150, 155, 157, 161, $163^{(2)}$, 169, $172^{(3)}$, 174

Robertsons Fork (Meherrin River), 27
Miery Branch, 150
Miery Branch (low Grounds of Cub Creek), 76
Miery Creek, 15, 16
Miles's Creek, 9, 12[(2)], 17, 28, 60, 63, 78, 89, 90, 91, 119, 155
Mittle Creek, 153
Mountain/Montain Creek, 8, 65, 73, 121, 125, 131, 164
Musterfield Branch, 53[(2)]
North River, 10, 11[(2)], 17, 18, 19
Notaway/Nottoway/Notway River, 6, 10, 11, 33, 37, 43, 44, 48, 50, 67, 79, 118, 120, 121, 124, 128, 145, 149, 156[(2)], 164, 167
Ochenechee, 18
Otter River, 6[(3)], 8, 14, 23, 24, 35[(3)], 39, 47, 55, 64
ffish Dam on Otter River, 14, 23, 35, 47, 55, 65
great Otter River, 55
Little Otter River, 39, 55[(2)]
great Owls Creek, 44[(2)], 96, 108, 112[(2)], 165
Piney Ponds, 111
Poplar Spring, 6[(2)], 24, 47, 65
Ready branch, 44
Ready/Reedy Creek, 17[(2)], 24, 41, 53[(2)], 70, 71, 92, 93, 100, 117, 121, 122, 134, 135, 142, 143[(2)], 150, 158, 172
Ronoke/Roan Oak/Roanoake River, 9, 13, 24, 29, 49, 53[(2)], 64[(2)], 81, 88, 91, 98[(2)], 99, 113, 119, 130, 141, 145, 169
Little Roanoake/Roenoke River/Creek, 5, 6[(2)], 7, 17[(2)], 18, 21[(2)], 24, 28, 37, 40, 44, 47, 51, 52, 58[(3)], 59, 62, 65, 68, 74, 76, 79[(2)], 84[(2)], 85, 95[(2)], 97, 100, 102, 108, 109[(3)], 114[(2)], 115, 118[(2)], 120, 122, 124, 128, 131[(2)], 132, 133, 144, 151, 159
Great Lick on Little Roenoke, 6, 11, 18
Roberson/Robinson Creek, 44[(2)], 125
Sandy River, 10[(2)], 29[(2)], 38[(2)]
Forks of Seneca, Sincker Creek, 8, 19, 20, 29
Smiths River, 22[(2)], 37
Snow Creek, 6
South River, 18[(2)]
Stanton/Standton/Staunton River, 5, 8, 13, 14, 15, 22[(4)], 23[(2)], 30, 31, 33, 34, 35, 42, 47, 49, 57, 66, 69, 70, 74, 75, 80, 169[(2)]
Stokes's Mill Creek, 94
Stony Creek, 66, 129, 130
great Branch which runs into Stony Creek, 66
Swish Creek, 127[(2)]
Thistle Creek, 157, 158
Turnip Creek, 23, 30, 43, 114[(2)], 115, 143, 150
Turnys Creek, 13
Twittys Creek, 11, 60
Wards fork Creek, 21, 32, 34, 58, 60, 65, 74[(2)], 78, 90, 125, 132, 150, 171
Wood Pecker/Wood Picker Creek, 68, 97, 105, 153, 157

Miscellaneous

Albemarle County, 35
Amelia County, 33, 63, 156
Amelia County Court, 11, 25, 33, 48, 50, 79, 118, 120, 149
Bedford County Line, 94, 133, 139, 141
Brunswick/Brunsick County line, 13, 80, 82, 129, 151, 163, 166, 170$^{(2)}$, 171
The Carrolina line, 170, 173, 174
Causeway, 47
Charlotte Court House, later site of Variously referred to as "The new Town" (Dalstonburgh) and "the Magazine" during this period, 100, 101, 102, 107, 108, 109$^{(3)}$, 118, 120, 121$^{(2)}$, 124, 125, 131$^{(2)}$, 145, 156, 159, 162, 172
Clerks Office, 18, 25, 29, 34, 49, 66, 165
Council, 13
The Country Line, not to be confused with the County line. This is the line between Virginia and Carolina, 13, 29, 91, 92, 99, 118, 119, 153, 169
the County line, 15, 16, 33, 39, 53, 64$^{(2)}$, 87, 95, 98, 105, 128
the dividing line, 5
Diers, Dyors Pine, 17, 24
Ellidges Old Rase Paths, 148
Erskin(e)s Store, 160, 167
Euing's fence, 64, 65
General Assembly, 113
Governor, 13
Granville Court House, North Carolina, 14, 16, 39
Hallifax County, 169
Hallifax County Court, 66
Hound's Creek Race Paths, 38
George Jeffersons Store, 143
Langlys Store, 128
Lunenburg Court House, not the present site, 7$^{(2)}$, 8, 9, 10$^{(2)}$, 11$^{(2)}$, 12, 13$^{(2)}$, 14, 18$^{(2)}$, 26$^{(2)}$, 27, 32, 33, 36, 42, 52, 54, 55$^{(2)}$, 56$^{(3)}$, 59, 60$^{(2)}$, 63, 68, 72, 75, 78, 83, 87, 88, 92, 100, 101$^{(2)}$, 102, 116, 117$^{(2)}$, 120, 126, 127, 131$^{(2)}$, 144, 145, 147, 153, 154, 156, 158$^{(2)}$, 165, 168, 171, 172
Mathew Marable's store, 42
Mayo Settlement, 22$^{(4)}$
the Meadows, 23$^{(2)}$
great Meadow, 17
Round Meadow (on Stoney Creek), 129, 135
the Mine, 102, 104, 107
Nalls Shop, 159
Pankeys Store, 159
the Parsons Barn (on Kings Road), 131$^{(2)}$, 136, 144, 145, 169, 172
The Point, 102, 106, 107

Posts of Directions, Sign Posts, 56[2], 58[2], 67[2], 79[4], 121[3], 159[5], 160
Prince Edward County, 78[2], 95
Prince Edward County line, 78, 106, 131, 142, 146, 147, 150
Clement Read's Office, 100
the School House, 161
Old School House, 145
the old Trap, 142, 143
Three Mile tree, 133
Well's Race Paths, 46
Woolf Pit near Kings Road, 12[2], 124
Richard Witton's Gates, 89
Thomas Wynns Old Muster field, 149

Roads

Road from Perrin Aldays to Thomas Vernons junior, 72

Allens Road, 40

Road from William Hills to Allens Creek, 132

Road from Allens Creek to Humphries, 164, 168

Road from Allens Creek to Miles Creek, 90

Road from John Ashworths to the fork of Allens Road, 40

road from Butchers Creek by John Potters to Allens Creek Bridge, 138, 140

Road from little Cheary Stone to Allens Creek, 26

Road from Allens Creek Bridge to Miles's Creek, 12, 63, 78

Allens Creek Church Road, 64

Road that Crosses Maherrin River at Mizes foard, from the County Line to Where the same intersects Allens Creek Church Road, 64

Road around the Head of Ash Camp, 96

Road from Ash Camp to Meherrin, 83

Lyddal Bacon's Roling way, 153

Robert Bakers Road, 19, 20

Road from John Phelp's Mill crossing Seneca a little below the fork, thence below the three forks of Falling River, thence below the forks of Cubb Creek to Robert Bakers Road, 19, 20[(2)], 29

Beard's Road, 6

Road from Otter River to the North end of the Long Mountain at Poplar Spring to Beards Road at the head of Appomattox River, 6

Henry Blagraves Roaling way, 163

Road from Blanks Ferry to Thomas Fosters, Fosters Ordinary, 125, 142

Road from Blanks's ferry to the Parsons Barn, 136

Road from Wilmoths Ordinary to Bollings Bridge, 159

Bollings Road, 160, 164

Road from Bryery to Bollings Road, 160

Road from the fork of Bollings Road to Willinghams Bridge, 164

Bookers Ferry Road, 143

William Bookers Road, 61

Road Cleared by William Booker along the Ridge between Crooked Creek & Bairs Element into the Road that leads by Jennings's Ordinary and from the Little Beaver Pond into the said Bookers Road, 61

Bouldins Path, 129

Bouldins Road, 37, 60, 122, 148

Road from Thomas Bouldins to Ash Camp, 83

Road from the old Road Near Thomas Nash's into the road at or near Thomas Bouldins, 68

Road from Samuel Perrins to Captain Bouldins, 41

Road from Twittys Creek to Mr. Thomas Bouldins, 60

Road from Godfrey Jones's Road into Bouldins Road, 148

Road from Cubb Creek Bridge to Bouldins Road, 37

Road from fishing Creek to John Boyds Ferry, 34

Road from Lawsons Mill to Boyds Ferry, 40

Road from Thomas Vernons to the Head of Bryery River, 78

Road from Breedloves Creek to George Moores, 110

Road from Breedloves Creek (fork of Juniper) to (great) Owls Creek, 44, 112

Road from the fork of Roberson Creek to Breedloves Creek, 44

Road from the mouth of Finneywood Creek to the fork of Roberson Creek, 44

road from Buckhorn to Delonys Old Ordinary, 64

Buffelo Road, 129

Road from Captain Hunts Road into the road at Gryme's foard thence into Buffelo Road, 129, 142

Road from Burgamy's Church to the Road that leads to Allen's Creek Bridge, 172

William Byrds Road, 59[(2)], 64

Road from Byrds Road to the Mouth of Butchers Creek, 59

Road from the County line near James Yanceys to William Roysters land on Ronoke thence into William Byrds Road, 64

David Caldwells Road, 90

Road from Francis Grahams ford on Cubb Creek to Wards Fork Bridge Commonly Called David Caldwells, Road 90

James Caldwells Path, 32

Callaways Road, 35

Road from the Foot of Johnsons Mountain into Callaways Road, 35

Callehams Road, 133

Cart Path that leads from Callehams Road to the Chappell, 133

road from the Three Mile tree to the Head of the Haw Branch to the Cart Path that leads from Callehams Road to the Chappell, 133

Road from Camps Ferry to Cross Road from Roysters ferry to Colbreaths, 144

Road from Camps Ferry to the fork of the Road to the School House, 161, 173

Cargills Road, 37, 65, 66, 91, 126, 153, 159

(Robertson's) Road from Cargill's Road to Cox's Road, 65, 66, 159

Cherry Stone Road, 33

Road from the Cherry Stone Road over Meherrin River at the Mouth of Little Beaver Pond to
Crooked Creek Bridge, 33

Road from Nathaniel Christians to Bedford County Line, 133

Church Road, 76, 78, 84, 90

Road from Ruffins old mill to the Church Road, 84

Road from the Cumberland Parish line to the Church on ash Camp, 152, 159, 160

Road from the old Road that Crosses Little Roanoake near Thomas Nashes to the Church, 40[2]

roads from Cornwall Parish Glebe to Sandy Creek Church and to the several Churches in the Parish, 170

Road from Kotnops [?] Road to Allens Creek by the new Church, 162

Road from the Church on Otter River to the said River to Clement Mobberlies to the South fork of little Otter River thence to Goose Creek near John Harvies, 39

Cobbs Road 13 Road from Mays's fford on Stanton River to Turnys Creek, thence to Cunninghams fford on Cubb Creek, thence down the Road ordered by Brunswick Court to Cobbs Road, 13

Cocks Road, 67, 97, 125, 137

Road from the North Maherrin into Abraham Cock's Road, 15

Coles Road, 23, 30, 58, 74, 85, 106, 121, 152

Road from low Ground on the North Side of Yards Fork Creek to Coles Road, 74

Road from Coles Road to Little Ronoke Bridge Near Clement Reads, 58, 85

Road from Meherrin River to washburnes Path at the forke of Coles Road, 121

Comers Road, 106

Road from Comers Road to the Prince Edward County Line, 106

Bridle way along the old Road from the County line to Langleys Store, 128

Road from County line Creek to fishing Creek, 33

Court house Road, 54, 60, 66, 78, 96, 97[2]

Road from Kings to Randolphs Road Commonly Call'd the Court House Road, 96, 97

Road from the Mouth of Blue Stone Creek into the Road Leading from this Courthouse to Twitty's Ordinary, 32, 33

Road from the Middle fork of Blew Stone Creek to the Courthouse, 60

road from the Court House to Great Blue Stone, 117

Road from the Court house to Blue Stone, 101, 102, 117

Road from the mouth of Butcher's Creek to the Court House, 126

road from the mouth of Butchers Creek to the Church and to the Court House, 165

Road from Byrds Road to the Courthouse, 59

Road from Cargills Ferry to the Court house, 8

Road leading from the North River at Cargill's Horse fford to the Courthouse, 11

Abraham Cooke's Road to be extended to the Courthouse, 11

Road from the Court house to Capt. Cox's at the fork of the same, 56

Road from the Court house to Delonys old Ordinary, 55

Road from the Court House to John Glasses, 159

Road from Courthouse into Hogins's Road, 7

Road from the Court house to the Horse Pen Creek, 72

Court house Road from the Horse Pen Creek to Bouldin's Road, 60

Road from the Court House to Humphris's Ordinary, 147

Road from the Courthouse to Lucas's Road, 13

road from the Court house to Mathew Marables, 55, 77, 87, 88, 116, 120, 156, 171

Road from Kings Road by Mathew. Marrables house to the Courthouse, 52

Road from Mathew Marable's Store to the Court House, 42

Road from William Marrables house into the Court house Road, 54

Road from William Marables Ordinary to the Courthouse, 36, 75

Road from the Court house to Marrables Ordinary, 63

Road from the Courthouse to Joseph Morton's Mill on Little Ronoke, 7

Road from Nottoway Road, tending across F****ing Creek, to Maherin Road, leading to the Court house, 18

Road from the Middle Fork of Maherrin River to the Court House, 26

Road around the plantation of Robert Munford and to the Court House, 145

old road leading from the Courthouse crossing Meherrin at George Elliotts House continuing
to Reedy Creek Church, 117

Road from Mitchells Ferry to the Courthouse, 8, 83

Road from Capt Mitchells Landing to the Court House, 27

Road from the Parsons Barn to the Courthouse, 131, 144

Road from Scott's fford on Maherrin River to the Courthouse, 10

Road from the South River to the Courthouse, 18

Road from Stanton River to the Courthouse, 33

Road from the Courthouse to Taylors Ferry on Roanoke, 168, 172

Road from Robert Wades to the Courthouse, 67

road from the Court House to Wade's Ferry, 127, 154

Road from Waltons Road to the Courthouse, 56

Road from the Court House to Colonel Wittons, 158

road from Colonel Harwoods ferry to the Road that goes from the Court House to Colonel Wittons, 158 road from the Mouth of Wood Pecker Creek into the road at this Court house, 68, 153, 157, 158

Road from the Wood Picker to the Mittle Creek and from the Kittle Creek to the Court House, 153

Road from the Wood Pecker to the Thistle Creek, 157

Road from the Thistle Creek to the Court House, 158

Road leading from the Courthouse to the Woolf Pit near Kings Road, 12

Road from Young's Mill to the Court House by way of Kings Road, 14

Cooks Courthouse Road, 18

Road from George Currie's Ferry into Cooks Courthouse Road, 18

Doctor Clack Courtneys Road, 164

Road from Cox's Bridge to the Middle Maherrin River Bridge, 163

Cox's Road, 37[(2)], 41, 50, 51, 65, 66, 124[(2)], 131, 133, 144, 159

Bridle Way from John Bacon's house to Cox Mill, 74

Road from John Cargills to Cox's Road, 37[(2)], 50

Road from Daniel Hayse's to great Notway on Cox's Road, 124

Road from the Hither Juniper to Cox's Road, 131

Road from Mathew Marrables Crossing the South fork of Maherrin River above & Near Millers Quarter thence into John Cox's Road, 51, 124

road from Cox's road towards Stoke Bridge, 133

Road from Wilmoths to Cox's, 145

Cubb Creek Road, 7, 87

Road from the Old Road to Cubb Creek, 146

Road from Cubb Creek half way to Prince Edward Line, 146

Road from the half way to Prince Edward Line, 147

Road from Cubb Creek to Little Roanoake, 37

Old road from the fork between Cub Creek and Stanton River Crossing the Old ford below Richard Dudgeons Plantation and into the road to Pattillo, 169

Bridle Way from the County line near William Watsons to Cubb Creek Road near John Wallers, 87

Road from Davis's ford to Claunch's Ordinary, 86

Delony's Road, 14

Road from Delony's Road to Jeffersons Ferry, thence towards Granville Court House, North Carolina, 14, 16

Dudgeons Road, 141

Dyers Road, 70, 90

Road from Bairs Ellement Creek to Dyers Road, 70, 90

Elledges Road, 48[(2)], 69[(2)], 70, 75

road from Tusling Quarter to Elledges Road, 70, 75

road from Ellidges road to Twittys Road, 69

Bridle way through the Land of George Elliot and John Robinson, 128

Road from Fall Creek to the Country Line, 35

Falling River Road, 5, 25, 122

Road from Marables Mill to falling River Road, 122

Road from the middle Fork of Little Roanoake into Falling River Road, 5

Road from Little Bair Creek to Falling River, 77

Road from Prince Edward County line to little Bar Creek, 78

Road from John Stewarts on Cubb Creek to the fforks of Falling River, 20

Road from the Ridge Joyning the County of Amelia to little falling River, 63

Road from the Forks of Seneca to the mouth of Falling River, 8

Road from Turnip Creek to Falling River, 23

Martin ffifers Road, 59

Road from the ford on Otter River Near the Mouth of Elk Creek down by William Callaway's Mill to the Ridge that Divide this County from Albemarle where Rusts Path Crosses the said Ridge, 35

Road from Christopher Irvins Ford in Otter River to William Calloway's Mill, 6

Roads from the Fish dam at Otter River to (James Johnsons) the Poplar Spring, 24, 47, 65

Road from the ffish Dam on Otter River to Goose Creek at a fford called Shorts fford, thence to Stanton River a little below Nicholas Rayle's, thence to Black Water, 14

New Road from Stanton River a little below Nicholas Hayle's to Black Water, 15

Road from the fish dam on Otter River to the Mouth of Goose Creek, 35

Road leading from the Fish Dam on Otter River to the Meadows, 23

Road leading from the Meadows to Maggotty Creek, 23

Road leading from Magotty Creek to the Burying Place at the End of the Road, 23

Road from Otter River at the Fish Dam Ford to Snow Creek, 6

Fish Dam Road, 55

Road from Otter River to the Fish Dam Road near the Church, 55

Road from Little Otter River to great Otter River, 55

Road from Goose Creek to Little Otter River, 55

Road from Goose Creek to the Extent of the County upwards, 55

Road from Otter River to Euings fence 64 Road from Euing's fence to the Blew ridge, 65

Road from flat Creek to Thomas Wells's, 54

flat Road, 15

Road from flat Road to the County line downwards, 15

road from Hust Brides (Hurt's or Hunt's Bridge?) to Flat Rock Church, 141, 147

Road from Francis Rays to John Jennings thence to flat Rock Church, 126

Bridle way from Ready Creek Church to flat Rock Church, 135

Road from Stokes Bridge to flat Rock Church, 130, 142

Road from Flatt rock Creek to Brunswick line, 170

Road from Flatt Rock Creek to great Creek Bridge, 121

road from flat Rock Creek to Brunswick line by Mason Bishops, 129

Road from Bears Element to the fork of the Road at flat Rock, 42

flat rock road 66, 83, 129, 130, 135, 155

Road from Gwinns Bridge to Young's Mill, 21

Road from Bouldins road crossing Little Ronoak at Gwins Bridge thence to Marables Mill, 122

Road from John Hites to Flatt rock Road, 155

road from the mouth of the great Branch which runs into Stony Creek to flat rock road, 66

road from Meherrin River near the mouth of flat Rock thence Crossing Stoney Creek and leading into Flat rock Road, 130

Road from the Round Meadow on Stoney Creek into flat rock road, 129, 135

Road from Fontains Ferry, 89

Fosters Road, 160, 162

Road from Richard Fox's Landing (on the South side of RoanoakRiver) to the Country line (towards Israel Robertsons Mill), 99, 153

Road from Louse Creek Bridge to Fuquays Ferry, 140, 141, 150

road out of Coles Road to Fuquays Ferry above Dudgeons Plantation, 152, 167

Road from Fuquays ferry to George Pattillo's, 167, 168, 169

Road from the Lower Falls of Banister River to Fuqua's Ford, 30

McNess Goodes Path, 88

Road from Graves ford to John Ashworths, 40

Road from Graves's fford to Scotts fford on Maherrin River, 12

the Great Road at Wilmoths ordinary, 165

Road from Joseph Greers to Mr Thomas Erskines, 158

Edward Harris's Path, 123

Hatchers Road, 81

Road from the Main Road near Hueys Cartpath towards Hawkins's fferry as far as the Country line, 29

Road from Hickmans Ferry to Miles's Creek Bridge, 9

Road from Stanton River to the Mayo Settlement at the Wart Mountains (later to be referred to
as Hickey's Road), 22 [4]

Initial Portions -
    Stanton River to Allens Creek, 22
    Allens Creek to Banister River, 22
    Banister River to Smiths River, 22
    Smith River to Wart Mountain Settlement, 22

Later Portions -
    Smiths River to Leatherwood, 37
    Leatherwood to the north fork of Sandy River, 38
    north fork of Sandy River to Bear Skin, 38
    Bear Skin to little Cherry Stone, 38

Road from William Bean's on Dan River to Banister, thence to the North River at Cargill's Horse Ford thence to the Courthouse, 10

Initial Portions -
    Beans to Sandy River, 10
    Sandy River to the Double Creeks, 10
    Double Creeks to Banister, 11
    Banister to the North River at Cargill's House fford, 11
    North River at Cargills Horse fford to the Courthouse, 11

Later Portions-

Double Creeks to the Middle fork of Miery Creek, 15
Middle fork of Miery Creek to Horse Ford on Banister, 16

Road from William Bean's House to the Head of Sandy River, 29

Road from the Head of Sandy River to the Lower falls of Banister River, 29

Hogins's Road, 7[(2)]

Road from the place where the new Road where of John Humphries is surveyor Intersects Hogins's Road to the Church, 7

Road from the Middle fork of Miery Creek to the Horse Ford on Banister River, 16

Road from Banister to the North River at Cargill's Horse fford, 11

Howards Road, 17[(2)], 28

Road from Howards Road crossing Miles's Creek, to Mizes Ford, 17

Road from Samuel Ashworths to Peter Hudsons foard on Stanton River, 57

Peter Hudsons Tract, 169

Hueys Cart Path, 29

Road from John Humphris's to Joseph Bozwells, 127

Captain Hunts Road, 129, 142

Hunts Ferry Road, 94

Road from James Hunts Ferry to Bedford County Line, 94

Road from Hunts Ferry to the County line, 105

Road from Hunts Ferry Road to Cubb Creek Bridge, 94

Road from Mays's Road to James Hunts Ferry, 82, 93

Ingrams Road (that goes over Roanoke River), 90, 98[(2)], 111, 145, 164

Road from Doctor Clack Courtneys Road to Ingrams Road upon the Ridge between Flat Creek and Dockerys Creek, 164

Road from Richard Fox's Landing on Roanoke River into Ingrams Road that goes over Meherrin River, 98[(2)]

Road from Ingrams Road to Mize's Ford Road, 90, 91

Road from the Piney Ponds to Inghrams Road, 111

Road from the Old School House into Inghrams Road, 145

Irish Road, 97, 137, 139, 149

Road from the Irish Road to the new Church, 149

road from Jones's Road into the Irish Road by George Fosters, 137, 139

road from where it crosses Island Creek round Daniel Mitchells Plantation, 155, 156

Bridle Way from Pinkethman Hawkins to Jeffersons Church, 110

Jeffersons Ferry Road, 78[(2)], 90

road from Jeffersons ferry to the Church (Road), 72, 90

Road from Jeffersons Ferry to the County Line, 53

Road from George Jeffersons Store by John Speeds to Cozens's, 143

Captain Jennings Road, 165

Road from the ford over Meherrin River at Francis Rays, to Captn. John Jennings Ordinary, 83

Road to the Clerkes Office from Captain Jennings Road, 165

Johnsons Road, 89, 154

Johnsons Road extended to Waltons Road, 89

Johnsons Roling road, 64

Isaac Johnson's Rolling road, 154

Road from Silvanus Stokes's Sr. on the Ridge between the Junaper and the Middle River of Maherrin down Johnsons Roling road to Silvanus Walkers Bridge at Scotts foard on the north fork of Maherrin River, 64

Johnsons and Waltons Road, 161

David Jones's Road, 25

Godfrey Jones's Road, 148

Godfrey Jones's Roling Road, 107, 108

Jones's Road, 106, 137, 139

Jones's new Road, 24

Road from Thomas Bouldins to Jones's new Road, 24

Alexander Joyces Road, 163

Road from the new Bridge on Cubb Creek to Alexander Joyces road, 163

Kemp's ferry Road, 154

Road from the Revd Mr. Keys Plantation, 94

Road from the Horsepen Creek to William Keys Plantation (the Parsons Barn), 94, 169

road from Claunches Ordinary to Kings ford, 88

road from King's Foard to the Country Line, 118

Road from that Road of Martin Fifers to Kings ford on Ronoke River, 49

Road from the Horse Ford on Ronoke to Kings Ford on the South side of Ronoke, 8

Road from Mitchells Ford to Kings Ford and thence to the Church, 9

King's Road, 7, $12^{(2)}$, $14^{(2)}$, 36, 38, 52, 56, 75, 78, $96^{(2)}$, $97^{(2)}$, 102, $103^{(2)}$, 117, $121^{(2)}$, 123, 124, 128, 131, 136, 137, 141, $149^{(2)}$, $154^{(2)}$, 168, 169, 172

road from Blue Stone to King's road, 117

road from blue Stone Bridge into Kings Road, 128

Road from Stony Hill by Cornelius Cargills to Kings Road, 96, 103, 154

road from Cocks Quarter on Roanoak to King's Road, 141

Road from Kings Road to the County line, 169

Road from Randolphs fork up to the Head of Reeses fork Where Edward Harris's Path Crosses Kings Road, 123

Road from Reases fork up to the Plantation of the Reverend William Kay, 123

Kings Road from Hudsons Foard to the Court House Road, 78

Road from Marables Ordinary to Kings Road, 36, 75

Road from Abraham Martins to Kings Road, 38, 103

Kings road from Martins Road to Robert Breedloves, 168

Kings Road from the Court House Road to Randolphs Road, 78

road from where Kings Road goes out of Randolphs Road to the Parsons, 121

Road from the Parsons Barn on Kings road to Martins Road, 172

Road from Kings Road to Wades Ferry, 136, 149

Road from Kings Road across Meherin River above Willinghams Bridge to Hampton Wades, 102

Road from Kings Road to Willinghams Road, 149

Road from Kings Road to Wimbish's house, 154

Road from the Wolf Pitt on Kings road to Joseph Williams's House thence to Kings road on Reeses fork, 124

Hugh Lawsons old rooling way, road along, 115

Bridle Way from Thomas Pound's up Ledbetter Creek to the road, 143

Lucas's Road, 12

Road from the Road that leads from the new Town over the County Bridge just above Clement Reads Plantation to the Bridge Built by him, thence to the Road down to the Court house and Church, 100, 101, 102, 109[(2)], 121

Road to Turn out of the Road near the new Town to Strike Godfrey Jones's Roling Road, 107, 108

Road from the New Town to the Road that Crosses Little Roanoak near Joseph Mortons, 109, 118, 119

Road from Boulding Church to the Magazine, 125

Road from Briery to the Magazine, 131

Road from the Magazine to Fosters Road, 162

Road over Mortons Bridge to the Magazine, 159

road from the Magazine to Roan oak bridge, 156

Road from the Magazine to the Bridge on Little Roanoke above Colo. Reads, 121, 124, 131

Road from Little Roanoake Church to the Magazine, 121, 131, 145, 172

Maherrin Road, 18

Road from Blew Stone to Maherrin below the ffork, 15

Road from Deloneys old ordinary over all the Branches of Meherrin into the Old road, 42

Road from Maherrin River to the Horsepen branch of Juniper Creek, 71

Road from ffork of Nottoway to Maherrin River in this County, 10

Road from Wittons fork over the North Meherrin to the old Road, 43

Road from the North fork of Meherrin River to the Middle Fork, 36, 41, 102

Road from Bares Element Creek to the Middle fork of Maherrin River, 26, 42

Road from Coxes to the Middle Maherrin River, 138, 161

Road from the middle Maherrin to Yilmoths Ordinary, 172

Main Road, 29, 153

Mr. Martins Road, 39, 44, 47, 62, 168, 172

Road from the Mossing ford (on Little Ronoke) into Martins Road, 39, 44, 47, 62

Martins Roling Road, 161

Bridle way from Thomas Bedfords Mill to Martins Roling Road, 161

Bridle way from Abraham Maury's House into the Road by Richard Elliott's Plantation, 170

Road from Daniel Mayse to the old Trap, 142

Mays's Road, 82

Mayse's ferry Road, 93[(2)]

Road from the Mouth of Elkhorn Creek to Mayses ford on Stanton River, 30

Road from Miles's Creek to Delonys Mill, 28

Road from Miles's Creek toward Maherrin, 155

Road from Miles's Creek Bridge to the extent of the County downwards, 12

Road from the Mine to Cocks Creek, 104

Road from the Mine to the Road that goes to the Point & etc., 102, 107

the Way From Mitchells Ford by Daniel Mitchells & Austin Spears's to the Country Line, 91, 92

Mitchells Road, 32, 80

Bridle Way from Pinkethman Hawkins's to Mitchells Road, 80

Road from Butchers Creek at Akins's Ford to Allens Creek into the Road that leads from Capt. Michells Ferry to the Point, 106, 151

Road from Colo Wittons to Mitchells Ferry, 96

Road from Aarons Creek to Robert Mitchells Ford, 16

Mize's Ford Road, 90, 139

Road from Mises Ford to Brunswick Line, 151

Road from Cocks Creek to Mize's Ford, 104

Road from Well's Race Paths by Delonys Ordinary into the Road at Mizes Ford, 46

Road from John Speeds into Mizes foard Road 138 Roads by the Plantations of John Speed and Sherwood Bugg, 171

Moores Road, 44, 108

Road from Bryery River to George Moores, 82

Road from George Moores to great Owl Creek, 102

Road from Owls Creek to Moores Road, 44, 108

Road from Owles Creek to Cross Maherrin to Mores below the old Road, 96, 108

Reuben Morgans Road, 54

Road from the Mossing ford Bridge to Captain Thomas Bouldins, Bouldins Path, 83, 113, 125, 129

Road from Thomas Comers to the Mossing Foard Bridge, 122

Bridle Way from the Mosing ford to the Reverend Mr. William Kay Deceased Plantation, 108

Road from Mountain Creek to Buckhorn and from the Mouth of the Creek to the extent of the County downwards, 8, 65, 73

Road from the Head of Montaine Creek to Nottaway River, 37

Nances Road, 6

Road from Nalls Shop to Pankeys Store, 159

Road from (Julius) Nichols's Ferry to the County Line to Meet the road leading from Granville Courthouse to William Eatons, 39, 98

Road from Dockerys to Julius Nichols's Ferry on Stanton Rive,r 31

Road from Brunswick line to Julius Nichols's Ferry, 80, 82

Julius Nichols's Road 90 Road from the North River to the South River, 18

Road from the head of Mountain Creek to Nottoway Bridge at Ralph Sheltons, 121, 131, 164

Road from Erskins Store to great Nottoway Bridge, 160, 167

Nottoway Road, 18

Road from Jennings's to Nottoway, 67

Road from Twittys to Jennings's, 67, 77

road from Aaron's Creek below Peter Overbey's to his Ferry, 120

Road from Palmers Ferry to Samuel Phelps's, 126

Road from Palmers landing in Butchers Creek to Martin ffifers Road, 59

Road from Reedy Creek Bridge to Diers Pine, 17, 24

Road from the North River to Reedy Creek Bridge, 17

Road from Reedy Creek to the Musterfield Branch, 53[2]

Road from Aarons Creek to Palmers ford, 57

Road from Palmers ford to Samuel Phelps, 126

Road from Mountain Creek to Parrishes fford, 125

Colo. Randolph's/Randolps/Randles Road, 5, 6, 12[2], 14, 21[2], 24, 26, 45, 46, 47[2], 56[2], 57, 59, 66, 67, 71, 78, 86, 92, 94, 96, 97[2], 103, 113, 115, 121, 131, 139, 141, 145, 147, 148

Road from the mouth of Ash Camp Creek into Colo. Randolph's Road, 5

Road from the head of Bear Creek into Randolphs Road, 147, 148

road from Bedford line to Randolphs Road, 139, 141

Road from the upper fork of Randolphs Road to Capt. Bouldins, 46

Road from the Court House to Randolphs Road, 92

Road from Randolphs Road (a little above George Moores) to the Mouth of Finny Wood, 27

Road from Randolphs Road to the Horse Pen (fork of Juniper Creek) Creek, 59, 71

Courthouse Road from Randolphs road to the Horse Pen Creek, 66

Road from Hudsons foard to Randolphs Road, 57

Road from Randolphs Road to Kings Road, 14, 56, 121

Randolphs Road from Thomas Worthys up to the Mossing foard, 21

Road from the Mossing Ford Bridge into Randolphs road, 86

Road from the fork of Randolphs Road to the Parsons Barn, 145

Road from Randolphs Road by Samuel Perrins, 26

Road from Randolphs Roaling Road to the Ridge Road that goes through Prince Edward County, 78, 95

road from little Roanoke Bridge to Randolph's Road, 115

Road leading from the Mouth of Little Roanoke to the Church along Randolphs Road, 21

road from Randolphs Road to Thomas's Ferry, 67

Road from Hampton Wades Road to Randolphs Road, 56

Road from John Young's Mill to Randolphs Road, 12, 14, 24

Road from Winninghams Ford on Maherin River to Randolphs Road on Little Roenoke as Twittys Path now goes, 6

Road from the fork of Randolphs Road to the Bridge over Maherrin River near Winninghams, 45, 47

Bridle Way from Willinghams road to Randolphs Road, 113

Road from the Woolf Pit near Kings Road to Randolphs Road, 12

Randolphs Roaling Road, 78, 95

Road from the Bridge to Randles Road, 47

Road from Roanoke river to the Fork of Randle's Road, 119

Randolphs fork, 123

Road from Dry Creek to Reedy Creek Church, 143

Road from the Glebe of Cumberland Parish to flat Rock and Reedy Creek Churches, 150, 158

Road from Reedy Creek Church to Daniel Hayse, 122, 134

Road from the North Meherrin to Reedy Creek Church, 142

Road from the old Trap to Reedy Creek Church, 143

Road from Willinghams Bridge on ledbetter Creek to Reedy Creek Church, 71

Reedy Creek Road, 172

Road from Barrs Element to the fork of Reedy Creek Road, 172

Ridge Road through Prince Edward County, 78

River Road, 5

River Road from the dividing Line up to Allen's Creek, 5

Road [River] from Allen's Creek to Butcher's Creek, 5, 34, 105

Road [River] from Butcher's Creek to Blew Stone, 5

Road [River] from Blew Stone to Cargills Ferry on Staunton River, 5

Road from the upper Bridge on Little Roanoke River, to Charles Andersons, 51

Road from the Middle Bridge over Little Roanoak River to the upper Bridge, 28

road from Little Roanoke Bridge to the Head of Bryery River, to the County, 95, 118

Road from the fork of Dunavant Creek to the Little Ronoke uper Bridge, 84

Road from Little Ronoake Bridge & Falling River, 95

Road from Prince Edward Line to Little Roan oak Bridge, 131

Road from Ruffins Quarter to Little Roanoak Bridge, 131

Road from the Ridge Road that leads from Rutledges ford to the Bridge over Little Ronoke, 17, 20

Road from Little Ronoke bridge to the low Grounds on the North side of Yards ford Creek, 74

Road from Little Ronoke Bridge to the head of the south branch of Yards ford Creek, 65

Road from Matthew Marables to Roanoke Church, 171

Road from George Waltons dwelling House to Little Roanoak Church, 92[(2)]

Roanoke Road, 44

James Roberts's Waggon Road, 52[(2)], 127

Robertson's Road, 65, 66

Road from the mouth of Aarons Creek to the Cross Road from Roysters Ferry to Colbreaths, 144

Road from William Roysters ferry to the Carrolina line, 170, 173, 174

Road from William Roysters Ferry to the Country Line, 119

Road from Roysters Ferry to the Court House Church, 131

Road from Grasty Creek to Roysters ferry, 130, 135

Ruffins road, 72

Road from Ruffins road to Bryery River, 72

Rusts Path, 35

Road from Saffolds ford to Brunswick Line near _____ Bridge, 163, 166, 170, 171

Road from Ellidges Old Rase Paths to Saffolds ford, 148

Road from John Humphris (Ordinary) to Saffolds foard on Meherrin River thence to flat Rock by Trumans Mill, 135, 174

Road from Saffolds ford on Meherin River to the Road at Capt John Jennings's, 83

Road from William Saffolds by Freemans Millon Flat Rock Creek to Kettlestick Creek, 49

Road from Coxes Ordinary to Scotts Bridge on Maherrin River, 172

Slate Rock Road, 41

Road from Reedy Creek to the fork of Slate Rock Road, 41

Road from Stewarts Ferry to where Thomas Comer lives, 121

Road from Stewarts Ferry on Stanton to Joseph Morton's Quarter, 70, 75

Road from Stewards (James Stewarts) Ferry on Stanton River to the Mossingford on Little Ronoke River, 47, 59, 84

Stewarts Ferry Road, 136

Bridle Way from John Owen's House into Stewarts Ferry Road, 136

Richard Stoke's Road, 56[2]

Road from Reedy Creek Church to Richard Stoke's Road, 56[2]

Road from Stokes's Bridge to the Road that Crosses William Crosses Bridge, 38

Stokes's Road, 67, 77

Edmund Taylor's Road, 154

Road from Allens Creek Bridge to Taylors Ferry, 172

Road from Taylors Ferry to Clarks Ordinary, 159

Road from Kemp's ferry Road over Butcher's Creek into Edmund Taylor's Road, 154

Treading Path (leading to Nichols's Ferry), 36

Road from Turnip Creek Bridge to Cub Creek Church, 150

Road from Turnip Creek Bridge to the new Church, 143

Twittys Path, 6

Road from John Davis's to Twittys Ferry, thence along the New Road to Granville C.H., N.C., 16

Road from Twittys Ferry to the County line, 16

Twittys (old) Road, 32[2], 41, 48, 52, 69, 112

road from Twittys old road to. John Hawkins's road, 112

Hampton Wades Road, 56

Road from Banister Bridge to Wades Ferry, 41

Waltons Road, 56, 71, 89, 125, 172

Road from the Robinson to Cocks Road called Waltons road, 125

Road from Cox's along Walton's Road to the Middle Maherrin, 172

Waltons Court House Road, 81

Bridle Way from James Roberts's Mill to Waltons Court House Road, 81

Road from where James Caldwells Path turns out of the Road to Wards fork Bridge, 32

Road from Wards fork Bridge to the new Bridge on Cubb Creek, 163

Washburnes Path (on Coles Road), 121

Williams's Road, 44

Road from Roanoke Road at the head of Ready branch into Williams's Road near the head of Crooked Creek, 44

Road from Cargills Road Opposite Cornelius Cargills to Willinghams Bridge on Maherrin River, 91

Road from the Bridge over Maherrin River Near Winninghams to the fork of the Road below Stokes's, 46

road from the road that crosses Blanks's ferry from William Tomasons into Willinghams Road near Mahering Bridges along the usual Tract opened by Peter Hudson, 169

Road from Winninghams Foard on Maherrin River to Nottaway River, 6

Willinghams, Winninghams Road, 11, 40, 52, 70, 81, 113[(2)], 127[(2)], 137, 149, 150, 160, 169

Road from the fork of Allens Creek to Winninghams Road, 40

road from Willinghams Road to Reedy Creek Church, 70

road from Willinghams Road to James Roberts's Mill, 81

Willinghams Road from Maherrin River to where James Roberts's Waggon Road comes into it, 52, 113

Winninghams Road from James Roberts's Waggon Road to the bridge near Hampton Wades, 52

Road from Willinghams Road to Wynn's Road, 39

Wisons Road, 81

Old Road from Wisons Road to Hatchers Road, 81

Witton's, Whitten's Road, 41, 43, 48, 49, 52, 67, 69, 83, 92, 93, 100

Road from Delonys Ordinary to Mitchells Foard & to the Fork of Wittons road, 43

Road from Mitchells Road about Half a Mile Below Henry Sages into a Road Made use of by Richard Wilton Gentleman Crossing Meherrin River below the Middle fork thence into Twittys Road, 32[(2)], 41

Road from Twittys Road where Wittons Road intersects the same into Eledges Road on the south Side of Maherrin River, 48, 69

Road from Ready Creek Church to the Forks of Wittons Road, 92, 93, 100

Wittons Road from the fork this side of Richard Wittons to the fork below North Meherrin, 83

Road by Grayors out of the Road by Richard Wittons, 132

road leading by Richard Wittons House to Gills ordinary, 134

Road from Colo Birds Mill to the Wood Pecker Creek, 97, 105

Wynn's Road, 39

Youngs Mill Road/Path, 16, 23

Road to Youngs Mill over Little Roanoke, 17

Road from David Logans to Youngs Mill Road, thence to Charles Talbots and to Mossing ford, 16

www.ingramcontent.com/pod-product-compliance
Lightning Source LLC
Chambersburg PA
CBHW060510300426
44112CB00017B/2619